High Mobility

A VISUAL HISTORY OF THE U.S. ARMY'S MODERN HIGH MOBILITY MULTIPURPOSE WHEELED VEHICLE, PART 1

by David Doyle

**M1038 • M1025 • M1044 • Avenger • M997 • M1151A1
M1165A1 • M1152A1 with B2 SECM**

Published by
Ampersand Group, Inc.
A HobbyLink Japan company
235 NE 6th Ave., Suite B
Delray Beach, FL 33483-5543
561-266-9686 • 561-266-9786 Fax
www.ampersandpubco.com • www.hlj.com

Acknowledgements:

This book would not have been possible without the generous assistance of Tom Kailbourn, John Adams-Graf, Wally and Nancy Wahlforth, Jeff Symanski and Randy Withrow. My wife Denise, who has an affinity for the HMMWV, consistently came through with both encouragement when needed, and the camera, ladder, or spare battery when those are needed—thank you Denise for all your help. All photos not otherwise noted, are credited to the author.

Sources:

Joint Mission Element Need Statement, Department of the Army, 14 February 1980

Acquisition Plan Revision for the High Mobility Multipurpose Wheeled Vehicle (HMMWV), Department of the Army, 14 September 1988.

Addendum to Joint Mission Element Need Statement, Department of the Army, 14 June 1989.

HMMWV M998A2 series Specifications and Performance Data, AM General, 2001

TM9-2320-280-10 Operator's Manual, Department of the Army, January 1996

TM9-2320-280-20 Technical Manual, Unit Maintenance, Department of the Army, January 1996

TM9-2320-280-24P Technical Manual, Unit Direct Support and General Support Maintenance, Repair Parts and Special Tools List, Department of the Army, March 2001.

TM9-2320-280-34 Technical Manual, Direct Support and General Support Maintenance, Department of the Army, January 1996.

Front cover: The ubiquitous High Mobility, Multipurpose Wheeled Vehicle (HMMWV), exemplified here by the M1165A1 of the 416th Theater Engineer Command, has come to symbolize American military presence. Originally conceived as an unarmored utilitarian tactical transport vehicle, the evolving nature of warfare has thrust the HMMWV into the role of armored combat vehicle. This M1165A1 typifies the latest generation of HMMWV. While the initial M998-series HMMWV weighed a modest 5,200-pounds, the weight of the M1165A1 has soared to a minimum of 7,230 pounds, with further increases of up to 1 1/4-tons depending upon optional armor configuration. The M1165A1 is categorized as an Expanded Capacity Vehicle and features a beefed up suspension and power train in order to accommodate the increased weight. ((Department of Defense)

Title page: The M998 and its winch-equipped sibling, the M1038, shown here, were the base vehicles for the first generation HMMWV. Subsequent generations of vehicles featured various armor packages, with ever-increasing weight, which in turn led to succession of engine upgrades. This M1038 is owned and preserved by Wally and Nancy Wahlforth.

Rear cover: M1151A1 up-armored HMMWVs of the 416th Theater Engineer Command roll along State Street, Chicago, Illinois as part of the 2014 Memorial Day Parade. Just two of the many different HMMWV armor and armament configurations are shown here. The vehicle in the foreground, in addition to an Integrated Armor Protection package, features a Gunner Protection Kit (GPK), while the second vehicle is equipped with the more advanced Objective Gunner's Protection Kit (OGPK). (U.S. Army photo by Sgt. 1st Class Michel Sauret)

©2015, Ampersand Publishing Group, Inc. All rights reserved. This publication may not be reproduced in part or in whole without written permission from the publisher, except in cases where quotations are needed for reviews.

Table of Contents

M1038	5
M1025	46
M1044	51
Avenger	55
M997	67
M1151A1	76
M1165A1	99
M1152A1 with B2 SECM	104
Field use	107

The Howie Belly Flopper was the pre-WWII predecessor to the progression of vehicles which evolved into the HMMWV. (Robert Norman collection)

Introduction

As of this writing, over thirty years have passed since the initial production contract for the High Mobility, Multipurpose Wheeled Vehicle, or HMMWV was awarded—with development of the vehicle consuming four years prior to that. During this time the HMMWV has evolved from a soft-skinned tactical utility vehicle to serve in a myriad of roles, and due to the exigencies of war even evolving into an armored combat vehicle. To date more than 40 models have been produced.

A rudimentary, much less comprehensive, look at the HMMWV family would require far more space than is available in a single Visual History series book. In this volume we will look in depth at selected variants - specifically the M1038, an early soft-skin base vehicle, the M1044 early armament carrier, and the M1151A1 with B1 armor kit, representing the recent trend to mold the HMMWV into an armored combat vehicle. Also represented in this volume is the Avenger surface-to-air missile system, arguably the most formidable armament placed on the HMMWV platform, as well as the life-saving M997 field ambulance and a contact maintenance unit borne by the M1152.

There are a variety of models of these vehicles, in part due to the Army's initially assigning a different M-number to winch-equipped models than non-winch vehicles. This practice was discontinued with the introduction of the A2 series. However, as the vehicle has remained in production for three decades, with continuous improvements, variations abound. Future volumes in this series will highlight other members of the HMMWV family.

The HMMWV was a further development of the light utility/reconnaissance truck that many feel can be traced to the pre-WWII Howie Belly Flopper. This small vehicle led to the WWII jeep, which in turn begat a succession of vehicles after WWII, including the M38, M38A1 and the M151 series of vehicles. Similarly, the Dodge tactical trucks evolved into the post-war M37. Seeking greater mobility, the U.S. Army tested a number of vehicles, even putting one of the more radical innovations into production as the M561 Gama Goat. None of these really gave the military what they wanted, either falling short in mobility, load capacity or, in the case of the M561, being very maintenance-intensive.

Thus began the competition for a new vehicle. After extensive competition, AM General was awarded the contract to build the High Mobility Multipurpose Wheeled Vehicle in March of 1983. It was intended to replace tactical vehicles in the 1/2 to 1 1/2 ton range, and as the name states, perform a variety of functions. Production HMMWVs began to be fielded in October 1985.

The hoods of the vehicles are fiberglass, and the bodies are made of aluminum, assembled using rivets and glue. The trucks have four-wheel double A-arm independent suspension. Disc brakes, inboard mounted near the differentials, are provided for each wheel.

In 1989 the Joint Mission Element Need Statement for the HMMWV was amended to include provision for the "heavy HMMWV." This vehicle had increased towed load and payload capacities, and was intended to serve as a prime mover for light howitzers, towed Vulcan systems, S250/S315 shelter carriers and the basis for the contact maintenance truck. The heavy HMMWV was type classified as standard as the M1097 in May 1992, and entered production in September of that year.

The initial series was powered by the General Motors 6.2-liter Diesel engines, which was superseded by a naturally aspirated 6.5-liter Diesel engine beginning with the M1097A1, the chassis that formed the basis for the A2 series vehicles. The latest versions of the HMMWV, based on the Expanded Capacity Vehicle, are powered by a turbosupercharged version of the 6.5-liter engine initially developed by General Motors and now produced by AM General subsidiary General Engine Products.

The Expanded Capacity Vehicle (ECV), sometimes known as the Expanded Capacity Humvee (ECH), compared to the M1097A2, has an increase of 700 pounds in payload capacity, rising to 5,100 pounds, and a reinforced frame to handle the increased burden. The base vehicle for the ECV family is the M1113.

The Expanded Capacity vehicle formed the basis for the up-armored HMMWV, a concept itself borne from immediate need in the 1990s rather than the initial planning of the HMMWV program in the 1970s. As the U.S. became involved in various conflicts around the globe, the HMMWV was increasingly pressed into roles that resulted in it becoming a combat vehicle. As a result of this, in 1994 the firm of O'Gara-Hess & Eisenhardt began development of the M1109, an armored version of the Humvee. With its armor protection, the M1109 is able to withstand 155mm artillery airbursts or a 12-pound anti-tank mine or direct fire from a 7.62-mm weapon-firing armor piercing ammunition.

O'Gara-Hess & Eisenhardt also later developed the M1114 which was based on the M1113 platform. The M1114 can be armed with an M2 or M60 machine gun, or a 40mm grenade launcher affixed to a ring mount on the roof of the HMMWV. The locking bi-fold door that closes the ring mount opening is designed such that when open, it forms armor protection for the gunner's back.

For use as base security vehicles, the United States Air Force purchased 71 similar vehicles, known as the M1116. Like the M1114, the M1116 is based on the M1097 Expanded Capacity Vehicle (ECV) base chassis HMMWV.

In 2004 AM General began producing their own armor-protected HMMWVs built on the M1113 chassis. The two models introduced were the M1151 armament carrier and M1152 cargo carrier/prime mover. The four-seat M1151 and two-seat M1152 models differ from the M1114 in that in addition to the integral armor, there is provision for additional armor kits to be installed in the field by the crew when the need arises. The M1151A1 with B1 kit has armor protection equal to that of the M1114. The A1 model designator represents the fact it has Integrated Armor Package (IAP) installed, commonly referred to as "Underbody." Further, the crew is secured to new blast-absorbing seats by a three-point belt system. The vehicle's air conditioning system has also been improved over that of the M1114.

The M561 Gama Goat, named for its proponent, Roger Gamount, was a 1 1/4-ton capacity 6x6 that was a marvel in both off-road capabilities and mechanical complexity. Troops appreciated its go-anywhere capabilities and loathed both the extensive maintenance required and the deafening noise levels. (TACOM LCMC History Office).

The M151A2 was the pinnacle of the development of a 1/4-ton 4x4 design developed by Ford in 1951. Though early models were plagued with handling problems, most of these were resolved in the A2 version introduced in 1969 and shown here.

Externally, the grill work for the air conditioner has been changed, distinguishing the variants.

Following up on the Add-On-Armor program begun in 2004, the Army initiated a Fragmentary Armor Kit program, which created several types of kits. The various kits are numbered according to purpose. Thus, kits for different vehicles can and do have the same number, although the components are different.

For the HMMWV, Frag kit 1 provided supplemental armor for the vehicle's A, B, and C pillars, and door frames, while kit 1A provided an overlay for the rocker panels.

Kit 2 protected wheel wells, kit 3 fuel tanks, and kit 4 vehicle floors. Kit 5, the Objective Frag Kit, was an armored door and rocker panel kit. Various versions of this kit were produced for the HMMWV. Notably, the Frag 5 door for the M1114 weighs 600 pounds.

Frag Kit 6 as developed for the Up-armored HMMWV was designed to defeat Explosively Formed Penetrators. This kit included additional protection for the Frag 1 and 2 areas, as well as new armored doors.

Frag Kit 7, as stated by Major Carlos Lago, Assistant Project Manager for ARMOR (HMMWV), introduced …"Overhead Cover (OHC), which provides the capability for the gunner to view the battlefield without compromising safety. Fitted with transparent armored glass, the OHC provides protection from the sun, while still allowing the gunner to maintain situational awareness. A removable roof applique is also included to shield occupants from overhead threats, as well as new Load Range "E" tire and wheel assemblies that will provide increased reliability and durability compared to the current tire and wheel assembly. The BAE developed Vehicle Emergency Egress (VEE) window will also be included. It provides additional egress options through the driver and passenger side windows. The new Armor Suspension Kit (ASK) will also be provided, this new suspension will increase durability and compensates for the additional weight placed on the vehicle."

While it is envisioned that the HMMWV will be replaced by the Joint Light Tactical Vehicle (JLTV), this process will take many years, meaning that the HMMWV will continue to be a staple of U.S. forces well into the future.

Top: Although proposals had been solicited from 61 firms, army records indicate only three companies submitted prototype HMMWVs for testing by the military. This example is one of the prototypes produced by Teledyne Continental, currently in the collection of the Veterans Memorial Museum in Huntsville, Alabama. (Veterans Memorial Museum) **Bottom:** Following the 1982 sale of Chrysler's Defense to General Dynamics, the firm submitted a prototype vehicle for consideration. This vehicle features an aluminum body built by Sheller-Globe. This vehicle is now in the collection of the Heartland Museum of Military Vehicles, Lexington, Nebraska.

M1038

The M1038 HMMWV cargo/troop carrier is similar to the M998 HMMWV cargo/troop carrier with the exception that the M1038 has a self-recovery winch, which also can be used to recover other types of vehicles of a similar size and weight. This example of a U.S. Marine Corps version of the M1038 has the deep-water fording kit, which enables the vehicle to pass through up to 60 inches of water in a river or body of water with a hard bed, as opposed to the normal 30-inch wading capability without the deep-water fording kit. The M1038 seats four and has space for carrying cargo as well.

Top left: The front bumper, a channel-shaped unit, is viewed from the left side, with the lifting shackles on the front in view. The bumper is attached to a stamped bumper-mounting bracket on each side; it is the part with three holes in the side. **Top right:** The left front end is displayed, showing the left composite light positioned in a recess in the front of the fender. On the side of the fender is an amber reflector and, to the rear of it, a clearance marker light, also with an amber-colored lens. **Above left:** A front view of the M1038 shows the oblong slot in the center of the front bumper to accommodate the winch cable. The cable is fitted with a hook that is secured to the right lifting shackle. The winch has a capacity of 6,000 pounds. On the right side of the windshield is the intake stack for the deep-water fording kit, which supplies air to the engine. Hidden from view is the exhaust stack of the deep-water fording kit, just forward of the left rear tire. Also in view are the recessed headlights flanking the seven oblong louvers for the radiator and oil cooler. **Above right:** A close-up photo shows the tow cable with tow hook attached passing through the oblong opening in the bumper. Behind the bumper is the electric-powered winch. The winch drum is visible directly behind the slot for the tow cable.

Top left: Behind the louvers in the front of the hood is a heavy-duty wire grille to protect the radiator and oil cooler, mounted to the rear of the grille and positioned at an angle, rather than vertically. Atop the hood is a louvered grille. **Top right:** The louvered, ballistic grille on top of the hood is seen in closer detail. A wire grille is mounted inside this grille, for extra protection of engine components from damage from foreign objects or explosive devices. An airlift bracket is on each side of the grill. **Above left:** A composite light and a blackout head light are mounted in contoured recesses on the left fender. The blackout light gives the driver just enough light to drive by, and the composite lights have a turn signal (amber lens) and, below it, a blackout marker light.
Above right: The contours of the hood and the positions of the round, amber reflector and clearance marker light on the side of the left fender are shown, as well as the positions of the left headlight, blackout headlight, and composite light on the front of the hood.

Engine Data 6.2 Liter used in early vehicles

Make	General Motors
Model	6.2 naturally aspirated
Number/Arrangement Of Cylinders	V8
Cubic Inch Displacement	378
Compression Ratio	21.5 to 1
Horsepower	150 @ 3600 RPM
Torque	257 @ 2000 RPM

Engine Data 6.5 liter used in M1097A2 family

Make	General Motors
Model	6.5 naturally aspirated
Number Of Cylinders	V8
Cubic Inch Displacement	396
Compression Ratio	20.2 to 1
Horsepower	160 @ 3200 RPM
Torque	290 @ 1700 RPM

Engine Data 6.5 liter used in Expanded Capacity Vehicle family

Make	General Engine Products
Model	6.5 turbosupercharged
Number Of Cylinders	V8
Cubic Inch Displacement	396
Compression Ratio	20.2 to 1
Horsepower	205 @ 3200 RPM
Torque	385 @ 1800 RPM

Top left: The left headlight assembly is viewed close-up. The sealed beam light is mounted in a housing, secured with the retainer ring that is visible around the light lens. To the right, a close view is provided of the front of the blackout headlight. **Above left:** On each side of the hood is a hood latch, the base of which is mounted on the cowl, and it engages a catch on the side of the hood. Protruding through the hood are the airlift brackets, to which slings are attached when delivering the vehicle by helicopter.

The M1026 is an armament carrier version of the HMMWV equipped with a winch and basic armor. Positioned on a 32-inch ring mount on the hardtop roof is an M2 .50-caliber machine gun with a night sight. The winch cable is visible on the bumper.

Top left: The open hood of an M1026, an armament-carrier HMMWV, is viewed from the left side, showing the wire grills and the light mounts. The hoods of HMMWVs are formed from a fiberglass-reinforced composite, while the body is composed of aluminum. **Top right:** In the foreground are the left splash shield / radiator side guard, the hood prop rod (center), and the left airlift bracket. On the opposite side of the splash shield is the radiator, which is mounted at about a 45-degree angle and has a "no step" sign. **Above left:** The engine compartment of the M1026 is tightly packed. Originally, HMMWVs had GM 6.2-liter, 150-horsepower liquid-cooled V8 Diesel engines, but in 1993, 6.5-liter, 160-horsepower Diesel engines were introduced to the A1 series of HMMWVs. **Above right:** The engine compartment of an M1042 is viewed from the left side, with the radiator to the left and the engine at the center. In the foreground with the "no step" sign is the windshield washer reservoir. Above the engine, with a "no step" sign, is the air horn.

Left: In a close-up of a well-weathered M1038 HMMWV from the left side, the brake master cylinder is at the bottom left, and the Prestolite 60-amp, 28-volt alternator is to the right. Various alternators are found in HMMWVs, including 60-amp, 100-amp, and 200-amp types. At the top is the upper radiator hose, behind which is visible the radiator fan and fan shroud. The hose at the center is the control-valve hose, part of the fan-drive mechanism. The top of the radiator is to the upper left.

Right: A broader view of the engine compartment is shown. The air horn on this example (upper right, with "no step" warning) is black. Attached to it with clamp bands is the elbow (top center). Next to the rear of the alternator is the front of the left valve cover. The upper radiator hose is attached to the water crossover, atop the front of the engine. The fan, fan shroud, and top of the radiator are to the upper left. One of the two v-belts is visible between the alternator and the fan.

Top left: Facing toward the rear of the engine compartment, the oil filler cap is right of center. Attached to the side of the oil filler tube is the Crankcase Depression Regulator valve hose. To the upper left is the surge tank. The elbow (left) leads to the air cleaner, out of the view to the left. **Top right:** The engine compartment of an M1038 is seen from the right side, showing the coolant surge tank (lower left) and, left to right, the elbow clamped to the air horn, the top front of the engine, serpentine drive belt, fan, fan shroud, and radiator. The alternator is at the top. **Above left:** An M1038 HMMWV chassis with the body removed serves to display features of the engine and its accessories. In view are the radiator (right), right exhaust manifold, and air horn (with tape over the opening). The diagonal tube houses the transmission oil dipstick. **Above right:** The right side of the oil cooler, mounted on the front of the radiator, is shown close-up. At the center is the right airlift bracket, below which is the upper right control arm, part of the front suspension assembly. In the background are the engine and drive shaft.

The engine and, to the right, the transmission, are viewed from the right side, with the left airlift bracket to the left. A clear view of the alternator is provided. Orange glow plugs are in the cylinder head. Adjacent to the rear of the valve cover is the engine oil dipstick.

The M1038 HMMWV chassis is viewed toward the rear, showing the muffler, the drive shaft next to the muffler, and the rear suspension. The frame is constructed of high-grade alloy steel and is of the full-box type, with five heavy-duty cross members at key points.

The transmission on the first generation of HMMWVs, including this M1038, was the General Motors turbo Hydramatic THM 400 three-speed automatic. The A2 versions of the HMMWV use the Hydramatic 4L80E four-speed automatic transmission.

Between the two rear cross members of the frame is the rear differential. The positioning of the rear and front differentials and drive shafts above the bottom of the frame provide these components with protection from damage from striking the ground.

Top left: To the rear of the transmission is the New Process model 242 two-speed, full-time four-wheel-drive transfer case, which features a lockable internal differential that provides for independent drive or locking of the front and rear half shafts so they receive equal power.
Above left: The right rear suspension is shown. The HMMWV's independent suspension is a long-travel, double-A-arm type with coil springs. Inside of each coil spring is a hydraulic double-acting shock absorber. **Right:** The left front suspension of an M1038 chassis is viewed from the front. The top of the coil spring rests in the spring seat, mounted on the side of the frame. The top of the shock absorber is also attached to the spring seat. The bottom of the coil spring rests on the lower A-arm, or control arm, and the bottom of the shock absorber is fastened to the shock mount bracket. Each half shaft (or, axle: the left front one is to the right of the coil spring) drives a geared-hub final drive housed in the knuckle. The geared hubs are made by AM General and feature spur gears with a 1.91-1 ratio.

The underside of an HMMWV is viewed from under its front end. In the foreground is the stabilizer bar of the front suspension, secured to the frame with two retainer straps. The exhaust pipe is to the side and rear of the rather rusty engine oil pan.

In a view of the underside of an M1038 from the rear, the rear differential is to the top. In the next bay forward are the muffler, left, and the fuel tank, right, and just beyond them is the transfer case. The rear axle has been removed from the transfer case and differential. The disk brakes can be seen on either side of the differential

The right front suspension is depicted, showing how the bottom of the coil spring rests on the lower A-arm. Below the rubber boot that covers the axle where it enters the knuckle is the tie rod, which acts on the steering arm, mounted on the knuckle, to steer the wheel.

Top left: In a view of the front left side of the frame and suspension, the rear of the left front tire is to the left, and the upper A-arm and its mounting bracket, attached to the side of the frame, are prominently in view, below which are the left tie rod and the drag link. **Above left:** Although early HMMWVs were mounted with Goodyear Wrangler RTII 35-12.50 all-terrain bias tires on 16.5-inch rims, since the early 1990s these vehicles have been equipped mainly with Goodyear 37-12.5 radial tires, such as these Wranglers. **Right:** The driver's rear-view mirror is mounted on a C-shaped tubular support fastened to brackets screwed to the side of the A-pillar, the structural frame to which the windshield is attached and that also provides an element of protection to the crew should the vehicle roll over. As a troop/cargo carrier version of the HMMWV, the M1038 lacks a hard top but can accommodate a canvas top. Such a top is installed here, secured to fasteners mounted on the A-pillar. The structure at the front of the cab and to the rear of the hood is the cowl.

Top left: The driver's seat has a canvas cover. To the lower left are the service brake pedal and accelerator pedal. At the center is the parking brake lever. On the left side of the dashboard is the rotary control for starting the engine, and, below it, the light controls. **Top right:** The parking brake lever is viewed close up. The small lever to the right of it is the transfer case lever, for selecting driving range. The larger lever to the right is the transmission shift lever. The cap protects a vehicular diagnostic-test receptacle. **Above left:** Above the starter switches are a wait-to-start light, which goes out when the engine is ready to be started, and a brake warning light to signal problems with the brake system. The gauge below the red placard indicates if a restriction exists in the air cleaner. **Above right:** The instrument panel is viewed close-up. The red warning placard pertains to the hand throttle, which is only partially visible below the lower left corner of the instrument panel. Above the speedometer is a high-beam light indicator.

In this driver's-eye view, to the left of the steering wheel is the directional-signal lever. The instrument panel contains a speedometer/odometer, engine oil pressure and coolant temperature gauges, voltmeter, fuel gauge, and defroster and heater controls.

Top left: A view of the front left corner of the driver's station in the cab shows details of the steering wheel, instrument panel, and door. The vertical channel on the door is the guide track for the window. The round knob is the lock for the window. **Above left:** The tray situated over the interior tunnel between the driver's seat and the companion seat (also called the commander's seat; background) in the M1038 HMMWV is a radio rack, below and forward of which is the engine access cover. **Right:** A door for a soft-top-equipped M1038 HMMWV is shown open, its part number stenciled near the bottom. The door is formed from a fabric skin fitted over a lightweight frame and is mounted on two hinges fastened to the cowl. The knob to the left is a latch handle. The window is made of soft plastic and is opened or closed with a zipper running along the sides and top of the window. This type of door was intended for relatively safe zones of operation, and would provide no protection against roadside bombs or projectiles. The ballistic doors introduced on later models would address this shortcoming.

The same area seen in the preceding page is viewed in this M1038A2 HMMWV, showing communications equipment installed on the electronics mount. Below the rear edge of the mount are holders for voice-communications handsets.

The rear half of the left side of an M1038 HMMWV equipped with a soft top is displayed, showing the installation of the raised exhaust of the deep-water fording kit. A flange on the elbow at the bottom of the stack is attached to a flange on the muffler, with a gasket between the two flanges serving to seal out water. A heat shield formed from expanded-steel mesh surrounds most of the stack. Each rear seat—the left one is shown here—is equipped with a cushioned seat and back and a seat belt. To the left is the B-pillar, and behind the seat is the C-pillar.

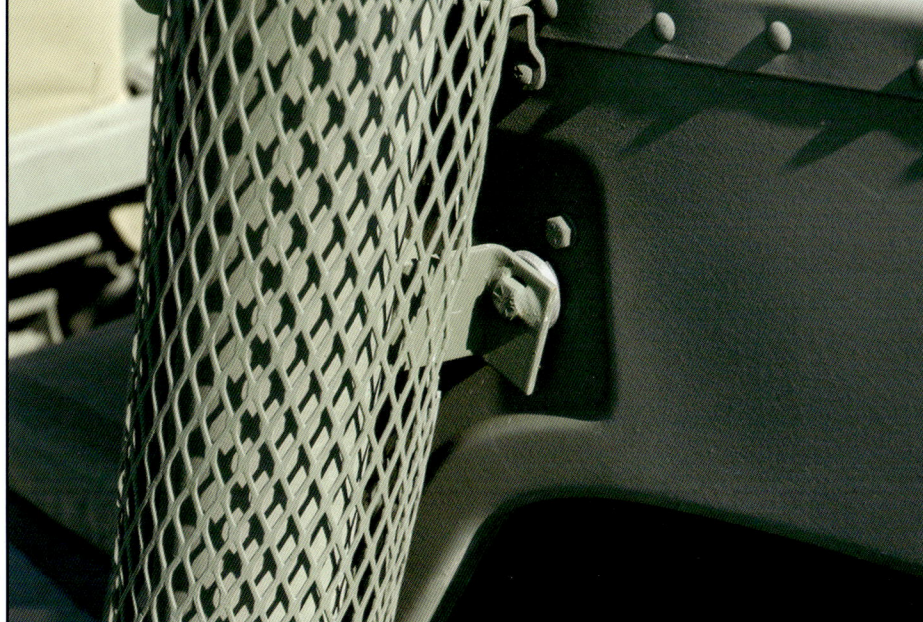

Top left: The deep-water fording kit exhaust assembly is secured to the side of the body of the vehicle with screws and nuts fastened to two tabs on the exhaust stack. This is the front tab. The expanded-steel mesh heat guard is also shown in detail. **Top right:** This is the rear tab securing the exhaust assembly of the deep-water fording kit to the side of the M1038 HMMWV. The area to the right is designated the left rear wheel house; in civilian terms, this translates to a wheel well. **Above left:** The interior of the cab of an M1038 HMMWV is observed from the left side. A soft top is installed. The B-pillar runs across the middle of the photo, and between it and the A-pillar is a detachable bow, to help support the canvas top. **Above right:** A detachable bow is also positioned between the B-pillar and the C-pillar (right) to support the canvas top. The bows are attached to horizontal rails between the pillars. Visible on the rear of the B-pillar are turn buttons, for fastening an optional curtain.

Top left: The left rear seat and seat belt are shown close-up. The interior tunnel running along the center of the cab is covered with an insulation blanket, and the floor of the cab is also insulated. The seat cushion is on the lid of a storage compartment with two latches. **Top right:** The left rear tire is a Goodyear Wrangler 37-12.50 radial, mounted on a 16.5-inch rim. The rim is secured to the wheel spindle with eight self-locking 9/16-18x1.50 studs. These are run-flat tires, with rubber inserts to allow them to keep running even when shot-up. **Above left:** The left rear wheel and tire have been removed from this M1038, exposing the knuckle, coil spring, upper A-arm, and some of the frame to view. Also in view is the lower part of the fording-kit exhaust, showing where its flanged end is joined to the muffler. **Above right:** On the left rear side of the body is a round reflector and, below it, a clearance marker light. Whereas the front side clearance marker lights have amber lenses, the rear clearance marker lights have red lenses. This is also the case with the reflectors.

Top left: Viewing inside the rear left wheel house toward the rear, the housing of the rear composite light is at the center of the photo. Below the lateral brace, to the lower right, is the housing of the left rear side clearance marker light. Electrical leads are visible. **Top right:** Underneath the rear of the HMMWV is a pioneer-tool tray holding an axe (visible here), shovel, mattock handle, and mattock head. Two straps are used to secure the tools to the rack, and releasing two latches allows the rack to swing down for access to the tools. **Above:** An M1038 HMMWV cargo/troop carrier is viewed from the side. It is fitted with a soft top over the cab, canvas-covered crew doors, and a brush guard on the front end. A canvas top is also installed over the cargo compartment in the rear.

The bottom extensions of the brush guard on this M1038 HMMWV are attached to brackets jutting from the front of the chassis frame with two fasteners per bracket. Several designs of detachable brush guards have been used on HMMWVs, including a type with heavy steel mesh welded to strategic points on the guard.

While some models of the HMMWV had no rear bumper per se, only two bumperettes on the rear cross member of the chassis frame, some vehicles, including the M1097-series vehicles, the M1123s, the M1025A2 and M1043A2 armament carriers, and certain other models, were furnished with full-width rear bumpers, as seen here. These were fitted with two tow rings, a tow pintle, and a receptacle for an electrical connection to a trailer.

An antenna base is mounted on the right rear corner of his M1038 HMMWV. The trailer receptacle is to the right of the tow pintle. Details of the tarpaulin and the rear curtain with a soft-plastic window also are available.

Fabric cargo covers have been produced in standard height, that can accommodate seated soldiers, and low profile versions such as this, more suitable as a cargo cover. The vinyl-coated covers for the HMMWV have been manufactured in Green 383, Tan 686, and three-color NATO camo, seen here. (John Adams-Graf)

A left-rear three-quarters view of an M1038 HMMWV with a deep-water fording kit displays features of the rear of the body, including the composite lights, reflectors, and tailgate. A towing pintle is mounted on the center of the rear bumper.

The tailgate eases the job of loading the vehicle. It has two hinges at the bottom and is secured with chains to locking brackets fastened to the body. The extensions on the top corners of the tailgate act as stops. Visible below the bumper is the pioneer tool rack.

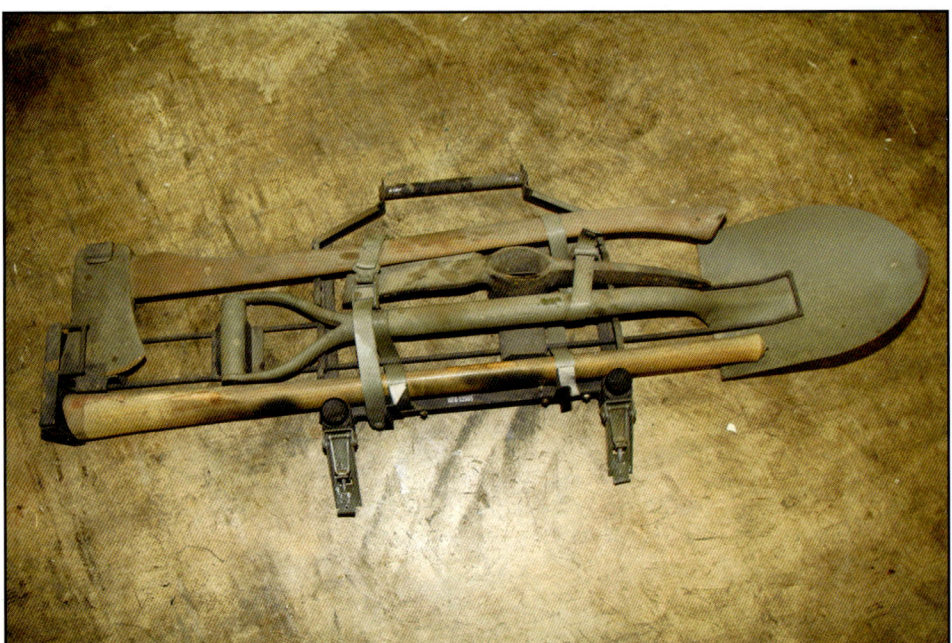

Top left: The two latches that hold the pioneer tool rack in the up position are visible to each side of the towing pintle, and the mattock handle is visible on the rack. Mounted above the right part of the bumper is a trailer-light receptacle. **Top right:** The red lens at the top of the rear composite lights is for a combination taillight, turn signal, and brake light. The rectangular strip immediately below it is the blackout marker light. The bottom rectangular lens is for the blackout stoplight. To the left is a reflector. **Above left:** On the right rear of the M1038 is a matching unit-base, or radio antenna mount, situated on an antenna-mounting bracket. Two curved guards on the lower part of the antenna-mounting bracket protect the electrical connections on the matching unit-base. **Above right:** The pioneer tool rack of an HMMWV has been removed from its position under the rear of the chassis frame. On the near side of the rack are the two latches. The rack holds a shovel, a mattock head, a mattock handle, and an axe, held in place by brackets and webbing straps. (James Alexander)

Top left: The lower A-arm and the lower part of the coil spring of the left rear suspension are in view. A good view is also provided of the non-directional tread pattern on the Goodyear Wrangler radial tires. The Wrangler bias tires on early HMMWVs had a different pattern. **Top right:** On the right rear of the M1038 HMMWV, the trailer light receptacle is above the bumper; the shovel blade is below the bumper. This type of narrow rear bumper was developed for Marine Corps HMMWVs; a wider type is used on U.S. Army versions. **Above left:** The pioneer tool rack is viewed underneath the M1038. In the bottom half of the photo is the rear cross member of the frame. Incorporated into the bottom corners of this cross member and another one forward of it are brackets for the lower A-arms. **Above right:** On the right side of the pioneer tool rack, the shovel, mattock head, and axe handle are shown. Below are the lower part of the rear cross member of the frame and the lower A-arm of the right rear suspension. To the bottom left, part of the rear differential is in view.

A USMC-type M1038 HMMWV cargo/troop carrier painted in a three-color camouflage scheme is viewed from the right rear. The tires have plenty of clearance in the wheel houses, or wheel wells, to accommodate the movements of the body and the wheels over rough terrain. In a round recess on the side of the body to the rear of the right rear crew door is the fuel filler cap. Protruding from the rear of the hood in front of the right rear-view mirror is the intake stack of the deep-water fording kit. This kit's exhaust stack is visible over the left rear of the vehicle. The deep-water fording kit is a stock feature of Marine HMMWVs.

Top left: A matching unit-base is shown close-up. A whip antenna would be attached to the top of the matching unit-base. The coil spring allows the whip antenna to flex when it hits obstructions. Also in view is the right rear composite light with its red lens. **Top right:** The matching unit-base is viewed from the left side. On its bottom surface are several terminals and connectors for attaching radio cables. The assembly is fastened to the top surface of the antenna-mounting bracket with several hex screws and nuts. **Above left:** The matching unit-base and the antenna-mounting bracket are seen from the right side of an M1038. The two small fixtures screwed to the top of the side of the body are footman loops, for securing straps attached to the bottom of a soft top. **Above right:** An HMMWV cargo/troop carrier is fitted with a soft top over the rear cargo bed. Four bows attached to the rear of the body support the canvas top. (John Adams-Graf)

The vehicle is configured for two crewmen, and the rear crew door has been covered with a blank piece of metal. The canvas front crew door and the soft tops over the cab and the cargo bed have been painted in a camouflage scheme, resulting in color tones that are somewhat different from those on the body of the vehicle. A brush guard has been installed on the front of the vehicle. (John Adams-Graf)

Top left: The cargo bed of an M1038 HMMWV is viewed facing toward the rear, showing details of the inner face of the tailgate. The design of the stamped recesses on the inner panel of the tailgate is different from the design of the outer panel of the tailgate. **Top right:** In the two-man configuration of the M1038, where the rear seats are removed, aluminum fillets are installed over the door openings. This one has a hinged panel at the bottom. To the left is the fuel filler cap; "use Diesel fuel only" is embossed on the rim of the recess. **Above left:** The cargo bed in an M1038 configured for a two-man crew is viewed from the right rear corner. For this configuration, a low bulkhead is installed at the front of the bed, and fillets cover the recesses for the rear seats. Stakes and troop seats are installed. **Above right:** The troop seats and stakes on the right side of the M1038 are shown. A support bar for the seat spans the gap between the wheel house (left) and the forward bulkhead. A curtain with a clear plastic window is fastened to the rear of the B-pillar.

The right side of an M1038 HMMWV cargo/troop carrier with a deep-water fording kit installed and configured for a four-man crew is displayed. The doors have been removed, showing the seats. Padding is present on the C-pillars and the rails on top of the A-, B-, and C-pillars, to protect crewmen should they jounce against these objects during travel over rough terrain. These pads are painted in the same camouflage colors as the rest of the vehicle. The bright green object at the bottom of the B-pillar is an informational placard.

Top left: The fuel-filler cap and its recess in the body are viewed. Embossed on the top and the bottom of the rim of the recess is "use Diesel fuel only." Above the recess is the bottom of the protective padding on the C-pillar. To the top left of the recess are two footman loops. **Top right:** The right rear crew seat is displayed close-up. Below the seat cushion is a stowage box with a hinged lid; one of its latches is visible. The back cushion is attached to the seat back via D-rings screwed to the seat back; one of these is also visible. **Above left:** The green placard next to the bottom of the B-pillar has instructions on installing a sling for airlifting, information on tie-down points, and weight and dimensional data, such as the shipping cubage of 666 cubic feet and the shipping weight, dry, of 5,187 pounds. **Above right:** The front right seat is designated the companion seat and, in some HMMWV configurations, the commander's seat. Its seat cushion rests on the hinged cover of the battery box. This box contains two batteries. Lying next to the battery box is the seat belt.

Top left: Above the interior tunnel, between the front seats, is the engine access cover, a molded assembly secured with two turn lock studs at the top and latches at the front lower corners and rear edge. To the far right are the heater duct and a data plate for the heater outlet. **Top right:** At the top of the center column of the windshield is a two-speed windshield washer motor, which drives a pivot shaft, which in turn drives the windshield washer arm. Electrical wires for powering the windshield washer motor are also in evidence. **Above left:** The crew heater is mounted on the firewall under the dashboard on the right side of the cab. Just below the windshield are adjustable baffles for blowing warm air. Warm air also blows out of the flexible hose. Heater controls are to the right of the instrument panel. **Above right:** Radio equipment is packed into the cab of this HMMWV, as viewed from the right side of the vehicle. Handsets are clipped to the two holders on the bottom of the radio rack, and a transmitter-receiver is on the rack. In the camouflage bag is a portable radio.

Top left: Unlike the left side of the cowl of the HMMWV, the right side of the cowl has five louvered openings in it. These are air intakes for the crew heater mounted inside the front of the cab. The design of the hood latch is also illustrated. **Top right:** A bracket near the top of the air-intake stack of the deep-water fording kit is fastened to the top of the A-pillar. At the top of the stack is a cap. Below the cowl, the stack is connected to the air cleaner, from which air is routed to the air-intake manifold. **Above left:** A thin tube, part of the deep-water fording kit, is attached to two brackets on the stack. A vertical seam is visible on the stack, as is the clamp for the cap. Also in view are the windshield wipers and two rubber bumpers on the top of the windshield frame. **Above right:** On HMMWVs not equipped with the deep-water fording kit, the air intake on top of the right side of the cowl is covered with a cap. The right rear corner of the hood (lower right of the photo) has a curve designed into it to fit around the air intake.

M1025

Several models of armaments-carrier HMMWVs have been produced over the years. The first-generation armaments carriers were the M1025, as seen here, and the M1026, similar to the M1025 but with a winch. The M1025 and M1026 have a hard top with a ring mount for a 40mm grenade launcher or an assortment of machine guns. Both models have a basic-armor package employing a mix of steel, Kevlar, and polycarbonate windows.

This HMMWV has a type of brush guard with heavy-duty wire mesh on the lower parts of the outboard sections and on the center section. This vehicle is equipped with a deep-water fording kit. On top of the right side of the cowl is the air intake, with a weather cap at the top of it. The M1025 can ford water with a hard bottom up to 30 inches deep, or 60 inches deep with the deep-water fording kit.

The M1025 can be differentiated from uparmored armament carriers by the X-shaped reinforcing embossed on the exterior of the doors. The uparmored armament carriers have smooth exterior surfaces on the doors. The hard top of the armament carriers consists of four main assemblies. The forward one is the roof, which contains the hatch and the weapons mount. To the rear of the roof are the sloping right and left cargo shells, between which is the cargo-shell door.

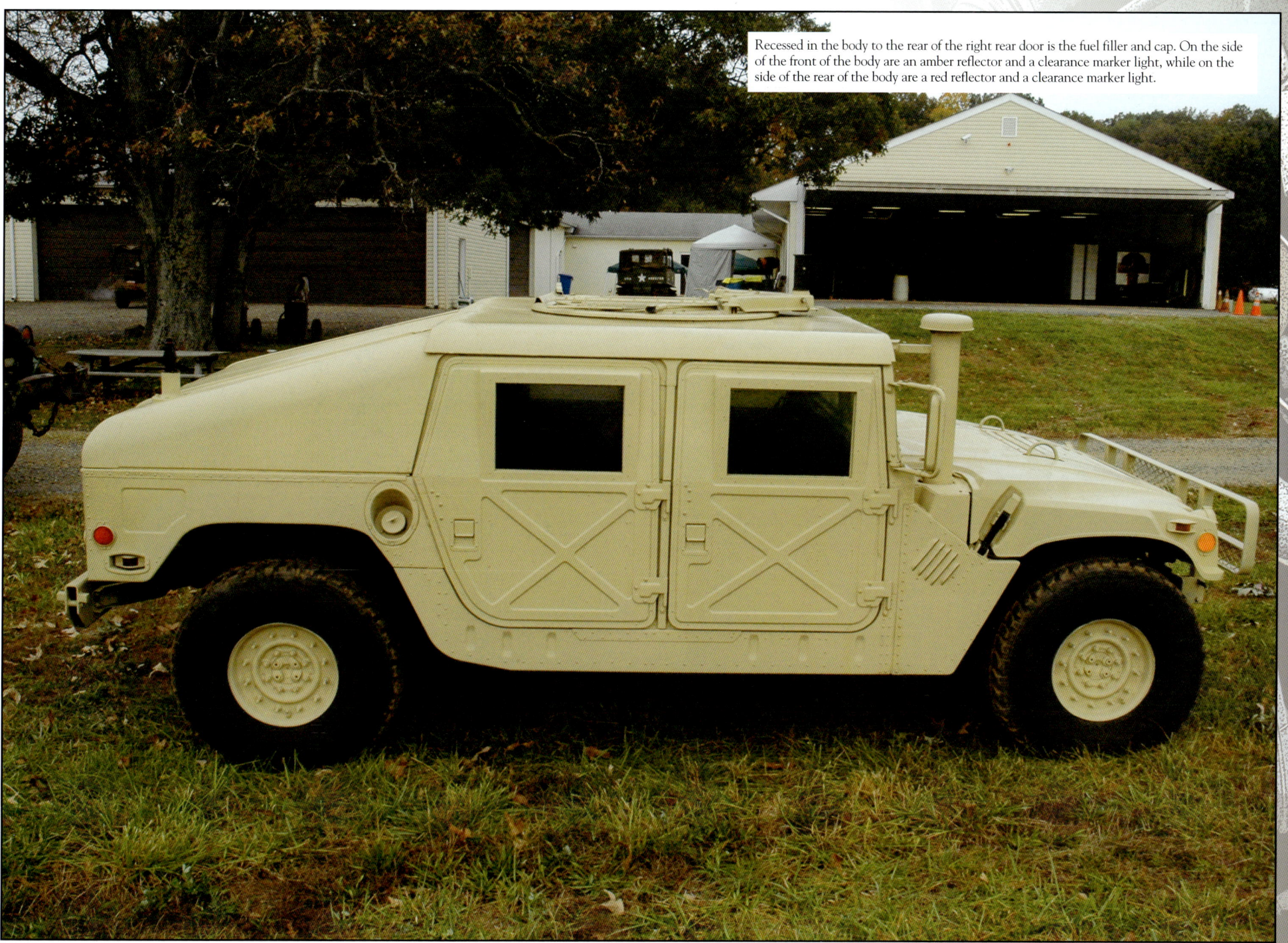

Recessed in the body to the rear of the right rear door is the fuel filler and cap. On the side of the front of the body are an amber reflector and a clearance marker light, while on the side of the rear of the body are a red reflector and a clearance marker light.

Top left: Another armament-carrier HMMWV is painted in a three-color camouflage scheme. A radio antenna and bracket are mounted on the rear of the left cargo shell. Note how the rear of the roof overlaps the forward upper edges of the cargo shells. (James Kettles) **Top right:** The armament-carrier versions of the HMMWV have on the roof a rotating structure called the turret. This is equipped with a hatch with a bi-fold cover, and a triangular plate designated the armament mount, visible here as the raised structure to the front of the hatch. A pedestal called the universal weapons adapter, not present here, was installed on the armament mount to provide a mounting base for the M60 7.62mm machine gun, M2 .50-caliber machine gun, or Mk. 19 40mm grenade launcher. **Above left:** An armament mount of a design with numerous lightening holes and a universal weapons adapter with a .50-caliber machine gun on it are seen from the front on a HMMWV armament carrier. Attached to the machine gun is a spotting scope. **Above right:** The same .50-caliber machine gun mount on an HMMWV armament carrier is viewed from the right side. The raised object on the near side of the turret is the turret-positioning handle, which provides the gunner with a positive right-hand grip for rotating the turret.

M1044

One of the automatic weapons the armament-carrier HMMWVs were designed to mount is the Mk. 19 40mm grenade launcher. The Mk. 19 has been in service since the Vietnam War. It is a fully-automatic, belt-fed, blowback-operated weapon that can fire at a rate of up to 375 rounds per minute, or a practical rate of fire of up to 60 rounds per minute. It has a maximum range of 7,257 feet. The Mk. 19 has a low recoil, making it suitable for mounting on the HMMWV, and it is effective against lightly armored vehicles and personnel. This example is mounted on an M1044 HMMWV in the collection of Steve Preston. (James Alexander)

The M1044 HMMWV is an armament carrier with supplemental armor and a winch produced for the U.S. Marine Corps. (The M1043 is a similar vehicle without a winch.) This M1044, viewed from above, is in the collection of Steve Preston. Mounted on the turret is a Mk. 19 40mm grenade launcher. A clear view is provided of the turret hatch with the hatch cover closed. (James Alexander)

Not so apparent in the preceding photos of Steve Preston's M1044 is the asymmetrical armored shield on the Mk. 19 40mm grenade launcher mount. The left side of the shield is designed to protect the ammunition box, whether the weapon mounted is a Mk. 19 launcher or a .50-caliber or 7.62mm machine gun. Also in view are two elements of the deep-water fording kit: the air intake on the right side of the cowl, and the extended exhaust tailpipe on the left side of the vehicle. This vehicle has the steel armored grille in the hood, both in the front and on top. The previous vehicles have the lightweight fiberglass grilles in them, which are not used on the soft skinned vehicles. (James Alexander)

Left: This M1044 was fitted with the prototype for the Warn 12,000-pound capacity front winch, which extended the front bumper forward. This winch was not adopted by the military. The standard M1044 winch was made by Mile Marker, and the bumper does not protrude. Also visible is the open top of the 40mm ammunition box on the left side of the Mk. 19 launcher. On each side of the receiver of the launcher is a charger handle. At the rear of the receiver are two grips, between which is a thumb-operated firing trigger. **Right:** The Mk. 19 40mm grenade launcher and its mount and shield are seen from the rear on an M1044 HMMWV. The launcher has two sights: a folding leaf sight on the upper rear of the receiver, and a blade-type front sight on the top of the top cover assembly. The round recess toward the rear of the right cargo shell is a mounting location for an AS-1729/VRC radio antenna; since the antenna is not installed, a protective plug is fastened in the recess. (James Alexander, both)

Avenger

Mounted on an HMMWV is an Avenger low-level air-defense system. Built by Boeing, it is composed of a fully rotating, gyro-stabilized turret, manned by a gunner with a standard vehicle mounted launcher (SVML) on each side containing four Raytheon Stinger short-range air-defense missiles. The pods have an elevation range of -10 degrees to +70 degrees. The system also includes a Belgian-manufactured M3P .50-caliber automatic machine gun, to cover the airspace the missiles cannot cover and to fire at ground targets. The HMMWV with the Avenger system is equipped with a hard top, and several versions of the vehicle, including the M998, M1037, and M1097, have been used as platforms for the Avenger system.

Another HMMWV with the Avenger system is viewed; the turret is traversed to the front, and part of the gunner's slanted windshield is visible between the Stinger launcher pods. The hardtop obviously is a necessity in an environment where missiles are launched.

Top left: The top of the Avenger turret, including the windshield, and the fronts of the Stinger missile launcher pods are visible above the vehicle's windshield. The turret windshield is hinged at the bottom, and the gunner enters or exits the turret through the opening.
Top right: Below the right Stinger pod (left in the photo) is a smaller pod containing the M3P .50-caliber machine gun. Below the left Stinger pod is a Raytheon AN/VLR-1 forward-looking-infrared (FLIR) unit used to acquire targets at night or when visibility is poor. **Above left:** The driver's door and side window and the left edge of the hard top are viewed close-up, along with the front of the left side of the turret. Stenciled on the side of the hard top is the part number and manufacturer's (Boeing) identification information. **Above right:** Affixed to a crossbeam supporting the Avenger turret is a data plate, listing the unit's serial number, contract number, part number, warranty code, NSN (national stock number), and more. The acronym in the "PMS/ TURRET MODULE ASSY" line refers to the weapon system's designation, Pedestal Mounted Stinger.

An Avenger low-level air-defense system is viewed from the rear. At the center is the rear of the turret, with a ventilation fan at the top center, flanked on each side by a matching unit-base with radio antenna (the right antenna is tied down). Other equipment on the rear of the turret includes the elevation drive motor and environmental control unit. Below the right Stinger pod is the .50-caliber machine gun pod; below that pod and next to the turret is a flex chute for .50-caliber ammunition. Below the left Stinger pod is the sensor pack.

Top left: The lower part of the rear of the Avenger turret is illustrated, along with a stowage box mounted on the rear of the body of the HMMWV. The box is constructed of diamond-tread plate and has two rear reflectors, two latches, and a hasp for a padlock. **Top right:** The hinged cover of the stowage box is open, showing the construction of the interior of the box. On the underside of the cover, D-shaped latch catches are visible. Two hold-open rods for the hinged cover are also present inside the box. **Above left:** The stowage box is not found on all HMMWVs with the Avenger. On the side of the box is a grab handle. Next to the box on the rear of the body of the HMMWV are a composite taillight and a reflector. On the end of the bumper is a shackle for an airlift sling. **Above right:** The right side of the stowage box is illustrated, showing the design of its handle and its diamond-tread plate construction. The box appears to have a bracket on its front side, which is secured with a pin with a retainer chain to the body of the vehicle.

Top left: Below the Stinger launching pod on the left side of the Avenger turret is the sensor pack, including the FLIR (Forward Looking Infrared). During clear, daylight conditions, the gunner uses a head-up CA-562 optical sight; in other conditions, he uses the FLIR to acquire and track targets. **Top right:** Components on the lower part of the rear of the turret are seen from the left. Because of the cramped confines of the interior of the turret, many systems and mechanisms are mounted externally, such as the elevation mechanism and air conditioner.

Above left: The Stinger launcher pods have upper and lower hinged and removable panels to provide access for loading the missiles. The Stinger, a "fire and forget" missile, has a range of approximately five miles and can attain an altitude of 10,000 feet. **Above right:** The sensor pack is viewed from a different angle. The FLIR allows the gunner to acquire and track a target in night or low-visibility conditions in conjunction with an automatic video tracker. Once the missile is fired, the gunner can immediately track another target.

Top left: The sensor pack is seen from the front. The equipment rests on a rack with a diagonal brace in front. In addition to the FLIR sensor and the automatic video tracker (AVT), the Avenger is equipped with an eye-safe laser rangefinder. **Above left:** The top of the Avenger is viewed from the rear. At the bottom is the cage for a ventilation fan. Non-skid strips are attached at intervals to the roof of the turret. On either side in the background are the tops of the Stinger pods. **Right:** Situated below the right Stinger pod, the Belgian-made M3P .50-caliber automatic machine gun is housed in a separate pod, with the barrel protruding. The gun is fed, using a flex chute, with ammunition from a box mounted on the right side of the turret to the front of the gun pod. This machine gun gives the Avenger a self-defense capability against enemy aircraft too close to engage with the Stinger missiles. It also provides the crew with a means of engaging enemy ground troops and light vehicles.

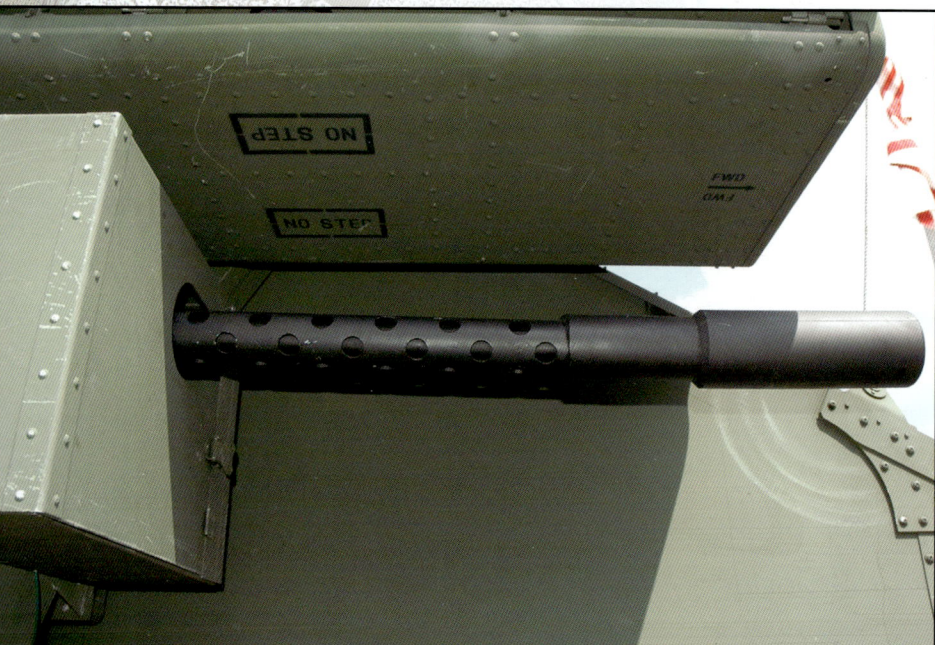
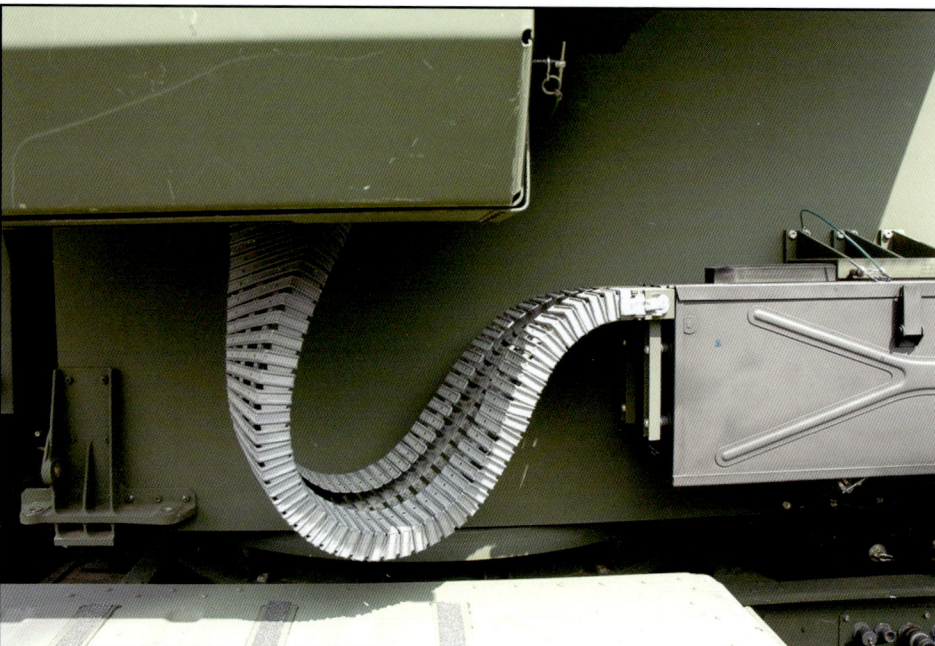

Top left: The Avenger system is viewed from the right rear quarter, showing the .50-caliber machine gun pod and the flex chute below it. The system is powered by its own batteries, which operate in parallel with the batteries and electrical system of the vehicle. **Top right:** The M3P .50-caliber machine gun and its pod are viewed from the front. Below the pod is the flex chute leading to the ammunition box at the bottom center; this box holds about 200 rounds of ammunition. To the top right is the gunner's window. **Above left:** The M3P .50-caliber machine gun is air cooled, and the barrel is fitted with a perforated cooling jacket, seen here. The fixture fitted over the muzzle of the gun appears to be a flash suppressor, to dampen telltale flames when the gun is fired at night. **Above right:** The flex chute for .50-caliber ammunition is viewed close-up. The ammunition is not loaded in this example. The ammunition box, to the lower right, is secured on a bracket on the side of the turret. More .50-caliber ammunition is stored in the HMMWV.

Top left: The .50-caliber ammunition box is shown in its mount on the lower front part of the right side of the turret. The M3P machine gun has a rate of fire of up to 1,100 rounds per minute; so frequent reloading of the box would be necessary in a combat situation. **Top right:** This is a close-up view of the rear of the right Stinger launcher pod on an Avenger, showing part of the inboard side of the launcher. The part number and manufacturer's code are lightly stenciled on the rear of the launcher. To the left is a whip antenna. **Above left:** The same whip antenna seen in the preceding photo is viewed from above, also showing the matching unit/base to which it is affixed. The yellow sign on it warns of the hazard of getting radio-frequency burns or shocks from the antenna. **Above right:** The bottom of the turret and its mounting stand on an Avenger on permanent display are shown from the rear of the HMMWV. A frame of channel-steel construction supports the stand. Note the absence of a tailgate on this and other HMMWVs with the Avenger system.

Top left: This view is looking into the well to the rear of the companion seat, with the right side of the frame supporting the Avenger mount running across the photo. Connectors for various electrical cables for the Avenger are mounted on the frame. **Top right:** The same section of Avenger mounting frame and electrical connections shown in the preceding photo is viewed from an angle. To the left is the right rear wheel house, and at the top left is the flex chute for the M3P .50-caliber machine gun. **Above left:** A hold-down cord for a radio antenna is wrapped around a bracket on the front of the Avenger turret. The slightly bulged panel to the front of the window is an identification friend or foe (IFF) sensor, designed to prevent the system from tracking friendly aircraft. **Above right:** Mounted in the cab of the Avenger between the driver's seat and the companion seat is the remote control unit (RCU). If necessary, the gunner can remotely operate the Stinger missiles from the cab using the RCU, or from up to 165 feet away from the vehicle.

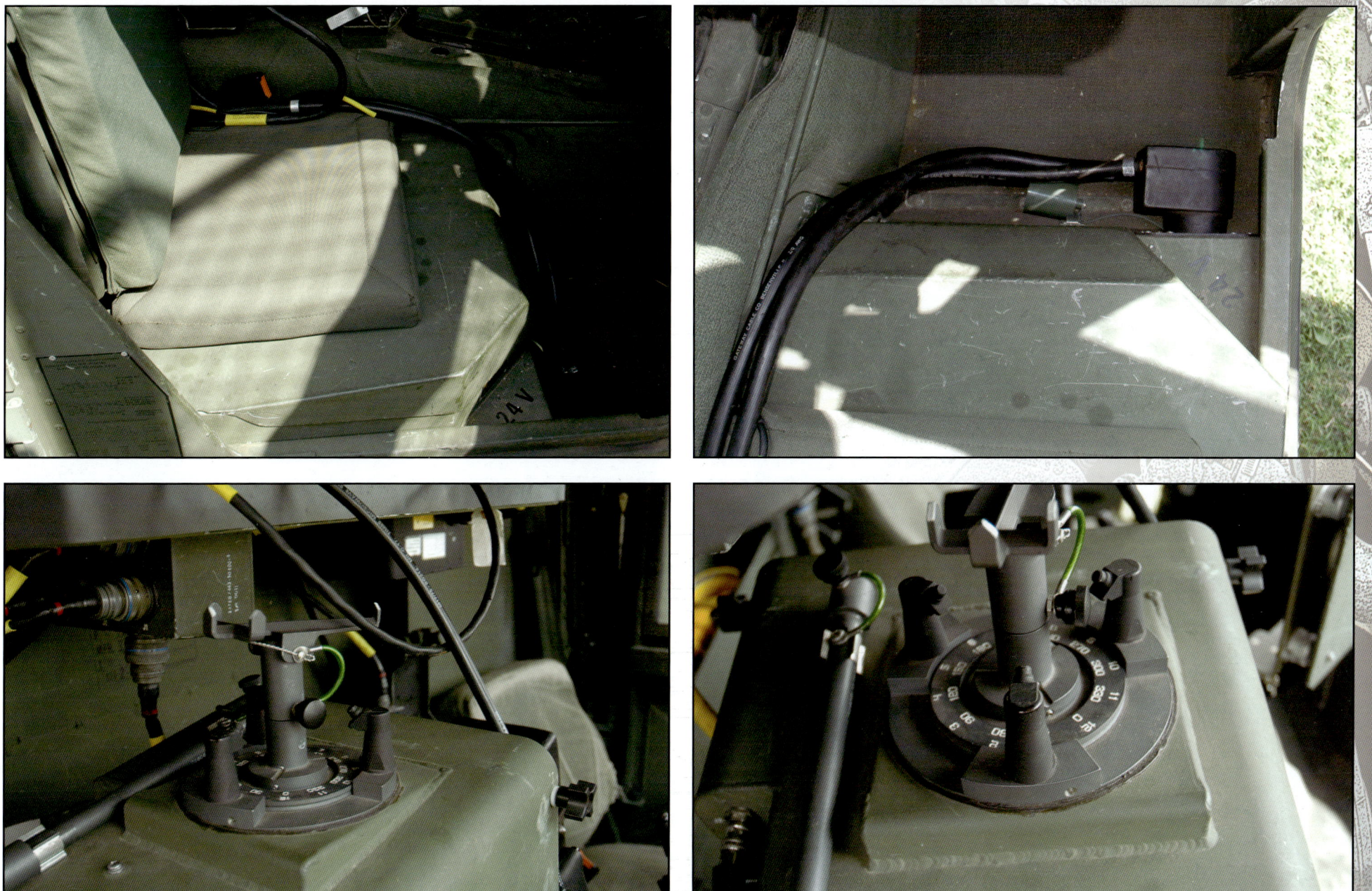

Top left: Electrical cables for the RCU are arrayed around the companion seat in the front right of the cab of an Avenger HMMWV. To the bottom left is the standard placard with weight and dimensional data of the vehicle and instructions for airlifting and for tie-down points. **Top right:** Cables are plugged into a receptacle on the front of the HMMWV's battery box, below the companion seat; the view is oriented with the front of the cab to the top. The HMMWV's electrical output rating, 24 volts, is marked next to the electrical plug. **Above left:** The top of the remote control unit is viewed. The instrument on top of it has a 360-degree scale around its base and apparently is used for taking bearings. In addition to firing the Stinger missiles remotely, the RCU can also be used to fire the .50-caliber machine gun. **Above right:** The remote control unit is seen from a slightly different angle. Upgrades to the Avenger's fire-control system now make it possible for it to link with other, more powerful, air-defense systems, such as long-range radar installations, thus enhancing its effectiveness.

Top left: The upper rear of the cab of an Avenger-equipped HMMWV is in view, showing various cables and electrical connections. Originally, the cab, like the turret, could get extremely hot during extended missions; so cooling systems were added to the cab and the turret. **Above left:** The Avenger system is equipped with a control box in the turret base, seen here with its cover removed. It includes missile-status indicators, a gun operation-mode switch, arming and laser-enable switches with red safety covers, operating-mode and turret-drive indicators, and circuit breakers. **Right:** The remote control unit (RCU) of the Avenger low-level air-defense system is viewed from the driver's side of the cab of the HMMWV. On the side of the unit is a keypad and display; to the sides of the keypad are grips. Switches are present on the body of the RCU for controlling the gun mode, weapons safety settings, power, monitor screen intensity, and more. The RCU is also equipped with a FLIR monitor, which displays the optical sight-generated reticles. When the FLIR is in tracking mode, it will show a tracking "box" around the target.

M997

The M997 is one of the Maxi-Ambulance family of HMMWVs, which also includes the M997A1 and M997A2. Other HMMWV-based ambulances included the M1035 soft-top ambulance and the M996 hardtop mini-ambulance, each of which could carry two litter patients or four ambulatory patients. The M997 has a higher profile than the M996, enabling attendants to stand up in the rear compartment. This vehicle can carry up to four patients on litters or eight ambulatory patients. It features a hardtop, air conditioning, and a basic armor kit.

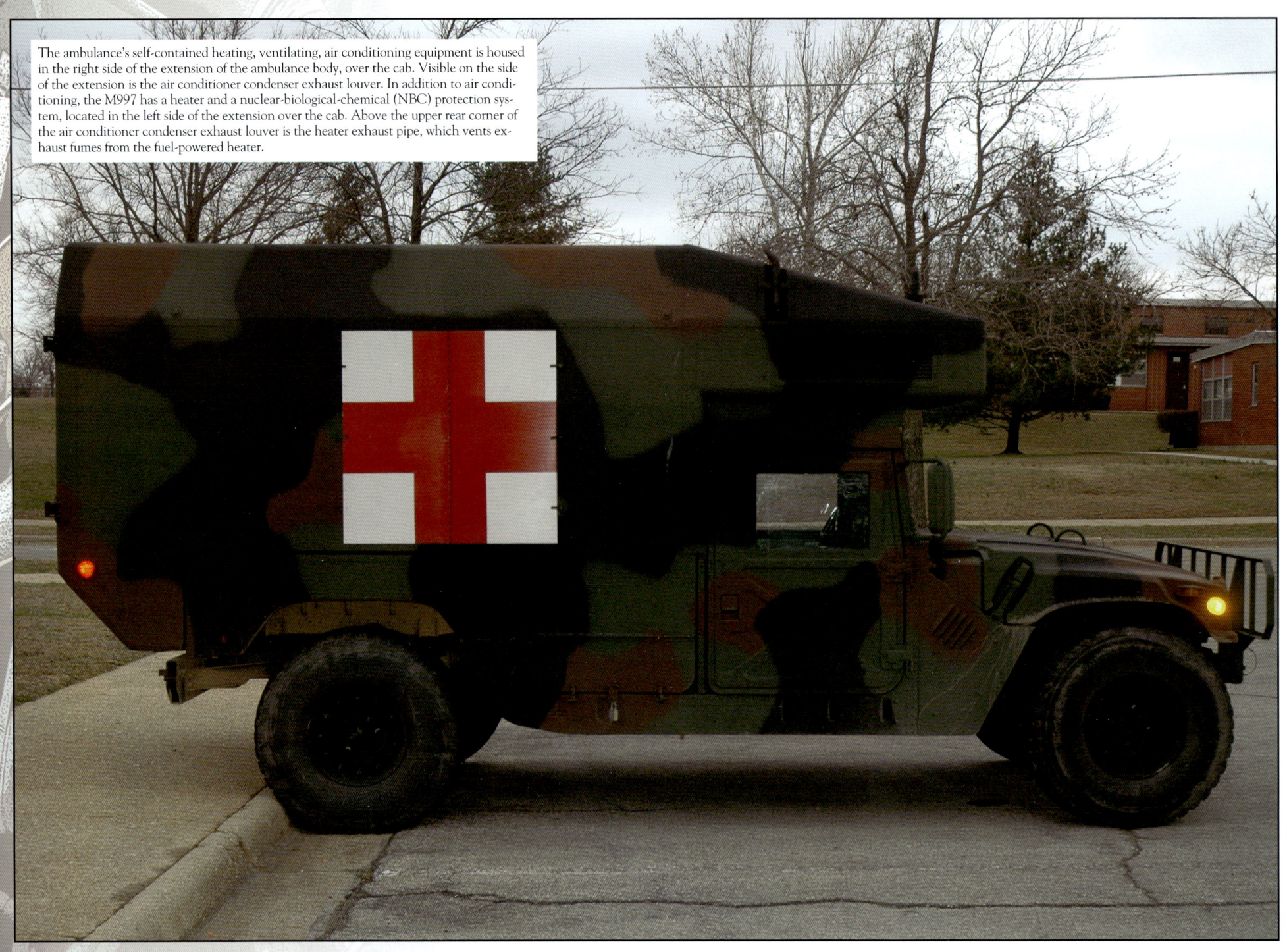

The ambulance's self-contained heating, ventilating, air conditioning equipment is housed in the right side of the extension of the ambulance body, over the cab. Visible on the side of the extension is the air conditioner condenser exhaust louver. In addition to air conditioning, the M997 has a heater and a nuclear-biological-chemical (NBC) protection system, located in the left side of the extension over the cab. Above the upper rear corner of the air conditioner condenser exhaust louver is the heater exhaust pipe, which vents exhaust fumes from the fuel-powered heater.

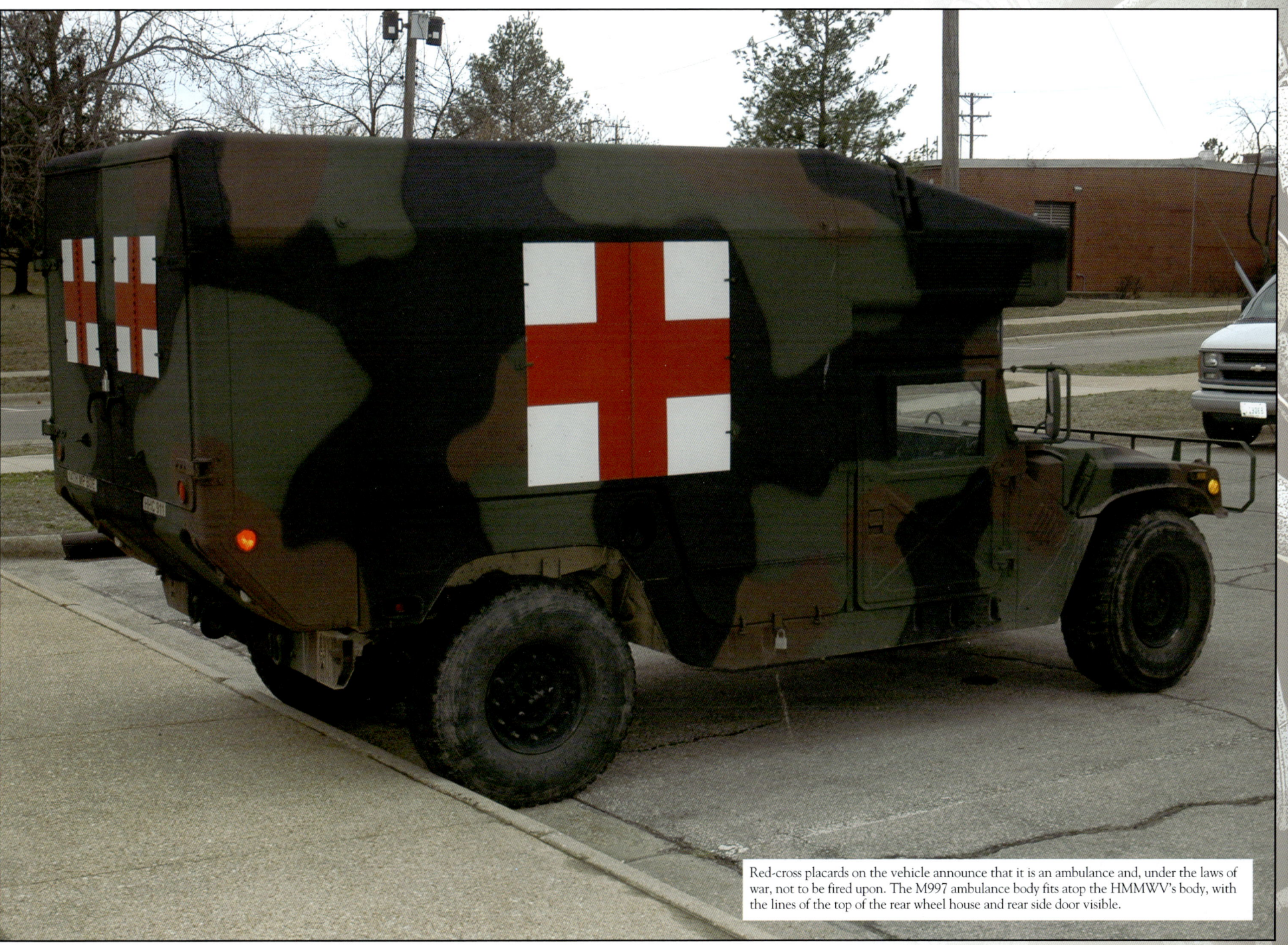

Red-cross placards on the vehicle announce that it is an ambulance and, under the laws of war, not to be fired upon. The M997 ambulance body fits atop the HMMWV's body, with the lines of the top of the rear wheel house and rear side door visible.

Left: The rear of an M997 Maxi-Ambulance is displayed, showing markings for the 14th Military Police Brigade. Below the twin rear entry doors is a folding step, shown closed. To the sides of the step are recesses containing rear composite lights and backup lights.

Right: The front of an M997 Maxi-Ambulance is shown. In addition to the red-cross placards on the sides, rear, and front of the ambulance body, another one is present on top of the ambulance body. The top of the heater exhaust pipe curves inward.

On the left rear of the ambulance body, between the folding step and the recess for the tail and backup lights, is a door to a compartment for storing litters and the litter rail extension. A latch for the door is above the door on the body.

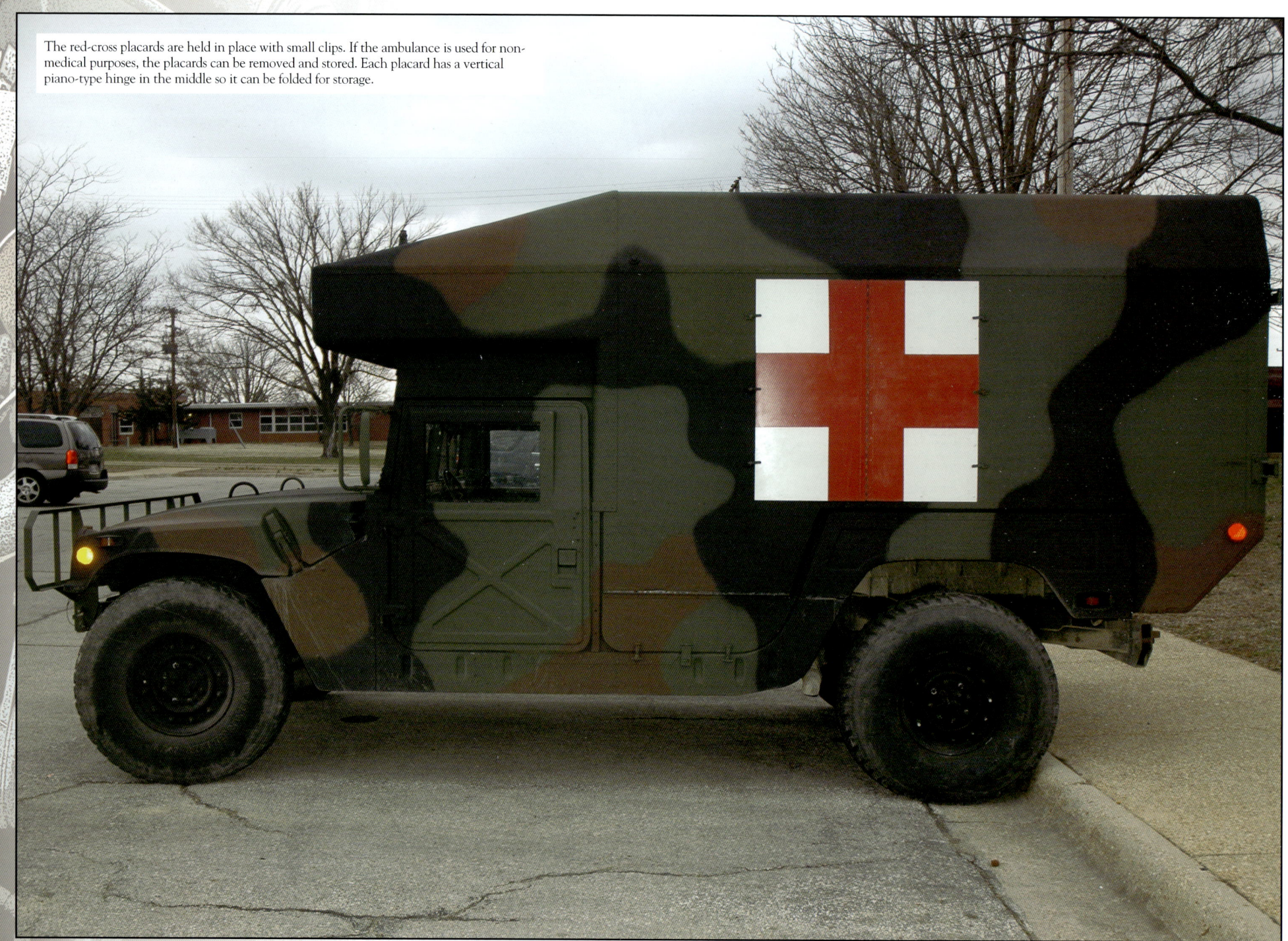
The red-cross placards are held in place with small clips. If the ambulance is used for non-medical purposes, the placards can be removed and stored. Each placard has a vertical piano-type hinge in the middle so it can be folded for storage.

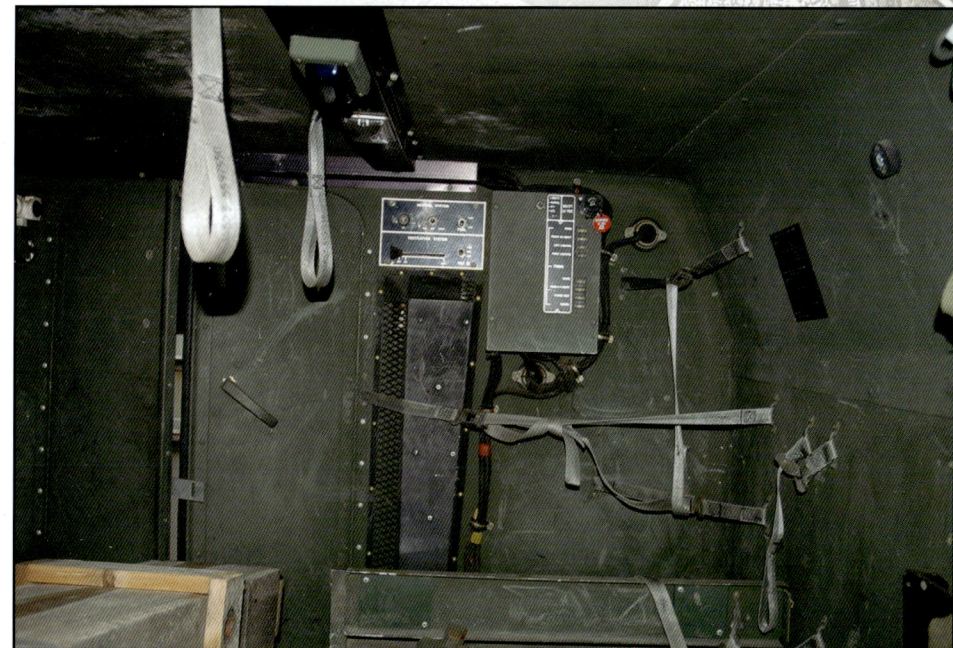

Top left: The M997 maxi-ambulance has built-in folding steps at the rear, with diamond-tread surfaces for non-slip use. To the left of the step opening is the door to the compartment for storing litter rails and rail extensions; the rail extensions are used for loading litters with patients onto the litter racks inside. Inside to the left is a dummy on a litter. **Top right:** In the left side of an M997, an upper litter rack and a lower litter rack are deployed. Both racks support litters with a patient. The upper rack is hinged to the body on the outboard side and supported by suspension straps on the inboard side. By releasing the straps and unlatching the upper rack, the rack can be lowered to form a backrest for ambulatory patients. To the right are bulkhead doors providing access to the cab. **Above left:** In the right side of an M997 maxi-ambulance, the upper litter rack is shown in the lowered position, to form a backrest for ambulatory patients. The lower litter racks are integral to the body and form a seat for ambulatory patients. Hanging over the upper litter rack/back rest are medical-supplies packs. (James Alexander, three) **Above right:** This M997 ambulance compartment lacks the partial bulkhead near the forward right corner. The black box mounted on the forward bulkhead is the control panel for the heating system and the ventilation system. To the right of it is the circuit-breaker box. At the top are two handhold straps and two ceiling lamps.

A brush guard is fitted on the front of the vehicle. On top of the extension of the ambulance body over the cab is a base mount for an AS-1729/VRC radio antenna. The grill on the right front of the extension is the air intake for the air conditioner.

An M997 ground ambulance stands ready for action during Vigilant Guard 2014, a multi-state emergency response exercise hosted by the state of Kansas. The 1077th Ground Ambulance Company out of Olathe, Kansas, had seven ambulances on hand both participating in the exercise and ready for real world emergencies. (Photo by Capt. Benjamin Gruver, 105th Mobile Public Affairs Detachment)

M1151A1

This is a factory-new M1151A1 w/B1 armor kit, an armament-carrier version of the expanded-capacity (ECV) HMMWVs, designed to handle a heavier capacity while maintaining the vehicle's dependability, mobility, and performance. The M1151A1 has the Integrated Armor Protection (IAP) package installed at the factory. This includes underbody armor, rocker armor, lower windscreen deflector armor, and energy-absorbing seats. This vehicle also has the B1 armor kit installed, with perimeter armor, overhead armor, and a rear ballistic bulkhead. On top of the cab roof is a gunner-protection kit (GPK). An air-conditioning system is also a feature of this model of HMMWV.

Top left: The tires on this M1151A1 w/B1 are B. F. Goodrich Baja T/A 37-12.5, load-range D, on 16.5" rims. Each rim assembly has 24 evenly spaced bolts. Should one or more tires be blown out, the vehicle has a flat-run capability of 30 miles at 30 miles per hour. **Top right:** Details of a portion of the underbody armor and side armor on an M1151A1 w/B1 are shown; the area is to the rear of the right front tire. Even the wheel house gets the armored treatment, with a panel visible directly to the rear of the tire. **Above left:** This view extends the view in the preceding photo, looking slightly upwards, and shows the armor treatment on the rear of the front right wheel house. A thin armor panel with notches on the top is welded to the larger armor plate below it. **Above right:** The armor panel at the center, on the right side of the cowl, is placed over the area normally occupied by louvered air intakes for the cab heater on earlier, unarmored HMMWVs. The hood is unarmored, and the hood latch is in view.

The M1151A1 w/B1 is equipped with ballistic crew doors with ballistic glass windows that jut out from the doors. The windshields are also of ballistic glass. The idea of such uparmored HMMWVs is to form a protective shell for the crewmen. The engines remain vulnerable with their fiberglass hoods. A deep-water fording engine-air intake is installed; it rises above the rear corner of the hood. On the roof of the cab is the gunner-protection kit (GPK), an armored enclosure that partially surrounds the weapons mount; the enclosure is seen here turned slightly to the left, and no weapon or frontal shield are mounted.

The front of the M1151A1 w/B1 presents a noticeably different appearance from the preceding HMMWVs in this book, with its box-shaped grille/headlight structure protruding from the front of the hood and the armored windshield. The "tie down" stencils below the service headlights refer to the shackles on the narrow front bumper. The opening at the front of the gunner-protection kit is visible; this enclosure is sometimes called a turret. In operational service, an additional shield is emplaced on the weapons mount, giving the gunner frontal protection.

Powering the M1151A1 w/B1 is the General Engine Products 6.5-liter V8 turbocharged Diesel engine, which produces 190 horsepower at 3,400 rpm. A four-speed automatic transmission is employed. To the sides of the oil cooler/radiator are airlift brackets. The reinforced windshield frame and ballistic-glass windshield, as compared to the basic, unarmored windshield of earlier HMMWVs, are evident. The windows on the M1151A1 w/B1 are formed of multi-ply bonded, ballistic-resistant glass. Several overhead armor panels are also in view on top of the cab roof.

The engine compartment is viewed from the left side of the vehicle. The device with the fins in the foreground is the dual-voltage regulator (14 volts and 28 volts), mounted on the alternator; over the rear of the alternator is a perforated fan guard. "No step" warning signs are placed on the top of the radiator, to the left, and on the engine air elbow to the upper center, to warn crewmen and maintenance personnel who are working on the vehicle with the hood open to avoid the temptation to plant their feet on these fragile, easily damaged objects.

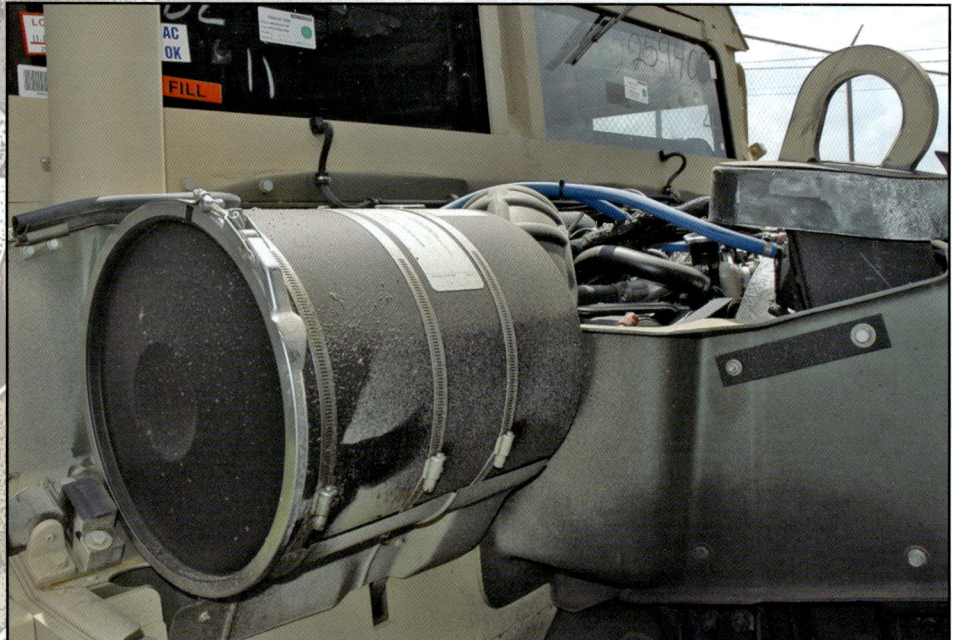

Top left: The drum-shaped air cleaner is secured to the front of the cowl with three clamps. The end of the air cleaner, to the left in the photo, can be removed by loosening the screw in its built-in clamp in order to inspect and clean the filter element inside the cleaner. **Top right:** On top of the air cleaner (bottom of photo) is a routing diagram for the serpentine belt. Next to the air cleaner is the surge tank, to the left. Under the blue hose, the ribbed fixture is the elbow that carries air from the air cleaner to the engine. **Above left:** The engine and oil cooler assembly is shown in the M1151A1 w/B1. As in previous models of the HMMWV, this assembly is positioned at about a 45-degree angle and is fastened to the top of the radiator assembly with four socket-head screws and washers. **Above right:** At the bottom center is the left splash shield of the engine compartment. The silver-colored fixture on the inboard side of the splash shield is the brake master cylinder. The brass-colored mechanism to the right is the power steering system hydraulic control valve.

The M1151A1 w/B1 is viewed from the left side. The extent of the armor protection is apparent, from the rocker panels to the doors and roof. The rear of the body, like the hood, remains unarmored. The crew doors are mounted on piano-type hinges. Mud flaps are installed on the rear bumper. The armor kit is removable, should it become necessary to convert the vehicle for other purposes where heavy armor would be a liability. Marked on the fenders are the recommended tire pressures: 40 psi for the front tires and 50 psi for the rear tires. The interior of the GPK enclosure on top of the roof is visible.

Top left: The right windshield and deep-water fording kit air intake are observed close-up. Various inspection stickers are applied to the ballistic glass, including a sticker identifying the glass panel as the right windshield and specifying its "threat side," or outer surface. **Top right:** The intricate patterns of the armor components on the front of the cab adjacent to the left windshield (to the left) are displayed. The armor panels and parts are fastened with hex screws and washers. Details of the side window enclosures are also shown. **Above left:** The blackout headlight (left) and front composite light are nestled in a recess on the front of the hood. The upper part of the composite light with the amber lens acts as a turn signal. The small rectangular lens below it is for a blackout marker light. **Above right:** Housed in the left side of the cargo shell (the hatch-back enclosure on the rear of the body) is the air-conditioning system. At the top of the wheel house is a ventilation grille for the air conditioner: this is a distinctive feature of the M1151A1 series.

A close-up view shows the add-on ballistic glass installation on the left rear door of a new M1151A1 w/B1. Unlike some of the earlier add-on ballistic glass assemblies on earlier HMMWVs, these units protrude very prominently from the doors.

Top left: On the armored doors of the M1151A1 w/B1 are heavy-duty D-rings, swivel-mounted on strong brackets. By attaching a chain or tow strap to the D-rings, another vehicle can forcibly pull off the doors if the crew should become trapped inside. **Above left:** At the top of each side of the B-pillar is a Y-shaped armored bracket that is fastened to the B-pillar and the roof armor with hex screws and washers. The use of screws on the armor allows it to be removed expeditiously if the vehicle is converted to peacetime use. **Right:** The two left-side windows are viewed in profile from next to the front of the cab. These multi-layered, bonded, ballistic-glass windows add a considerable amount of weight to the vehicle, and it was because the M1151A1 had received an improved, stronger suspension that such uparmoring was practical. Because it was necessary to operate in war zones with windows closed, it also became necessary to fit the M1151A1 w/B1 with an air-conditioning system, to spare the crewmembers from unbearable heat in warm or tropical climates.

The rear of the M1151A1 w/B1 is shown, illustrating the clamshell door that provides access to the interior of the rear cargo compartment and air conditioner equipment. Two radio antenna-matching unit-bases are present on the rear of the vehicle.

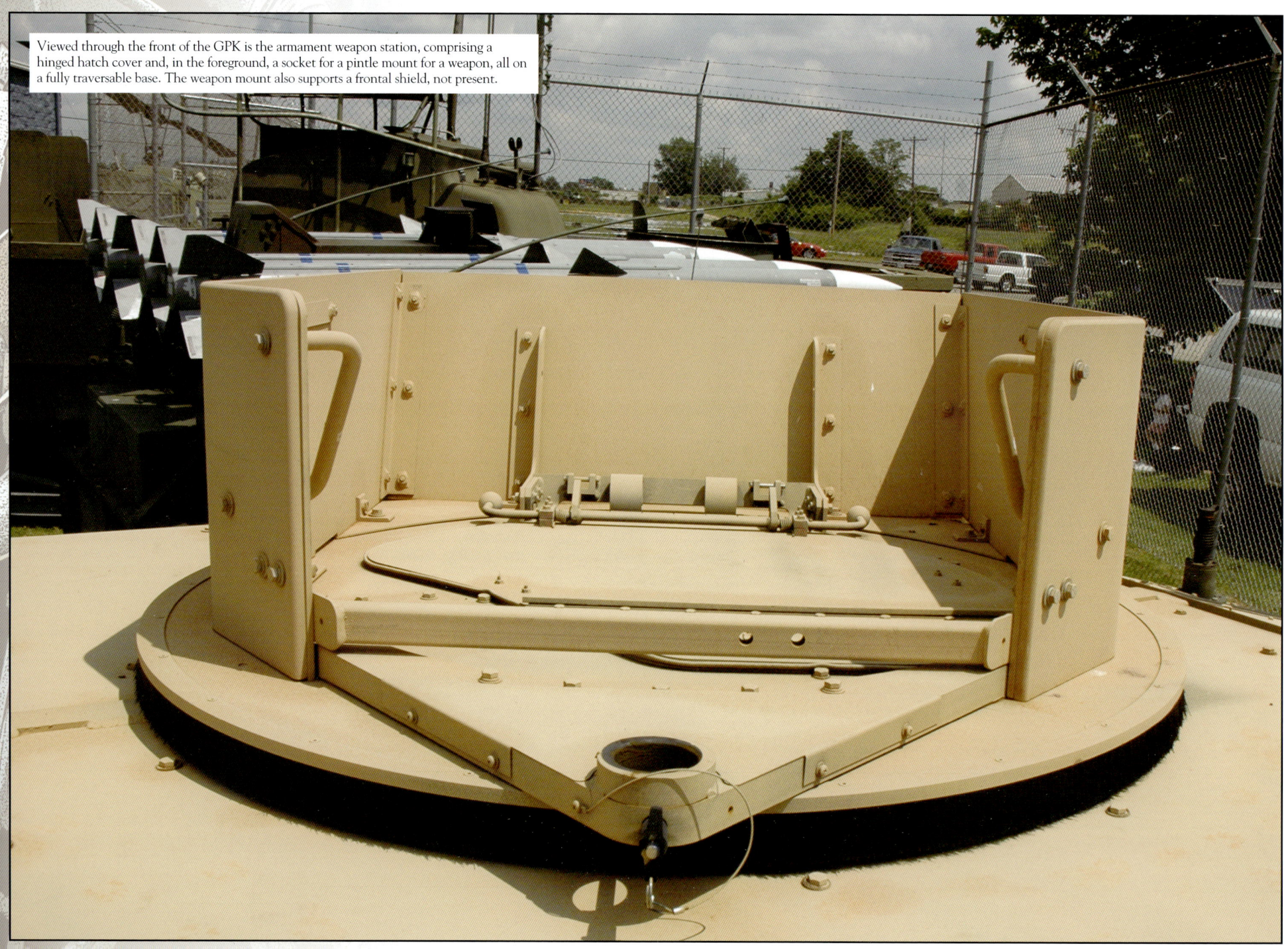

Viewed through the front of the GPK is the armament weapon station, comprising a hinged hatch cover and, in the foreground, a socket for a pintle mount for a weapon, all on a fully traversable base. The weapon mount also supports a frontal shield, not present.

The frontal shield and pintle mount are present on this armament weapon station, viewed from the left. The pintle mount accommodates the M2 .50-caliber machine gun, M60 or M240G 7.62mm machine guns, Mk. 19 automatic grenade launcher, and 5.56mm SAW.

The GPK gives the gunner a high degree of perimeter protection, but at the expense of visibility. Later, to provide better visibility, the Pickatinny Arsenal developed the Objective Gunner Protection Kit (OGPK), with integrated armor and ballistic glass.

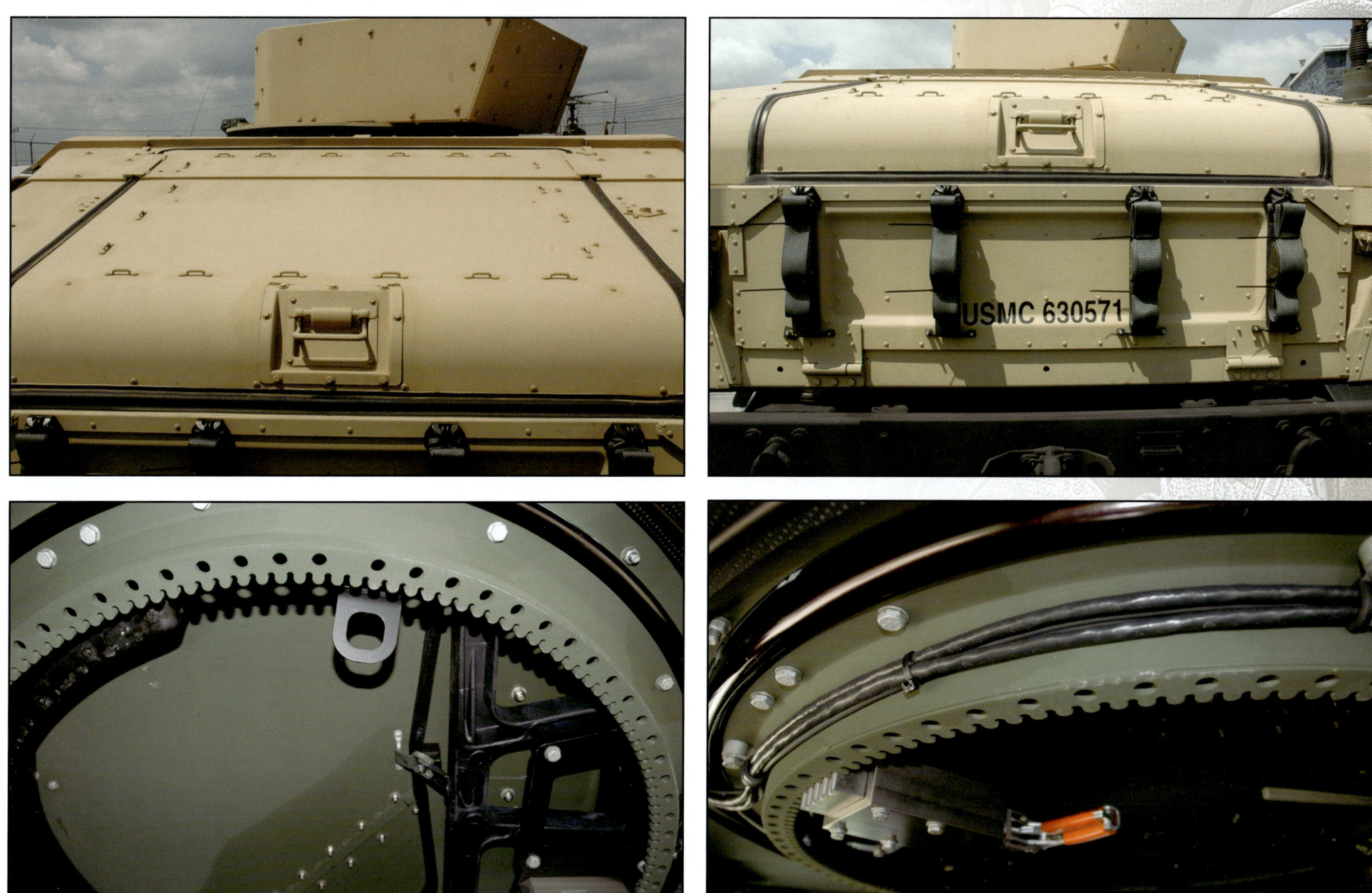

Top left: A closer look at the clamshell door is provided. Below it is the tailgate, secured with two hooks attached to chains. Webbing straps are attached to footman loops at the bottom and are hooked to brackets at the top, to secure a rolled camouflage screen. **Top right:** Arrayed on the clamshell door are footman loops: six on the top and bottom rows, and four on each side. These are used for securing a stowage net, which secures in place crew duffle bags. Also in view are the left side and rear of the GPK (Gunner's Protective Kit) atop the roof. **Above left:** When operating in the open armament weapon station, the gunner sits on a sling seat (not installed here) attached to two rings, one of which is seen toward the top. The gunner also wears a special harness that restrains him from being ejected from the vehicle. **Above right:** The underside of the armament weapon station is observed, showing the ring gear. Inside the ring gear, toward the left, is the turret brake. The orange handle on it is the turret brake lever. The gunner's body motion can be used to rotate the mount.

The interior of the armament weapon station is viewed from the front right (or companion) seat. On the back of the geared ring is a manual traversing unit, with a folding manual crank handle and an engage/disengage lever.

The dashboard and steering wheel of the M1151A1 w/B1 is similar in layout to earlier models of HMMWV. One difference is the air conditioner duct and vents on top of the dashboard. The left vent is to the upper left and the right one is to the upper right.

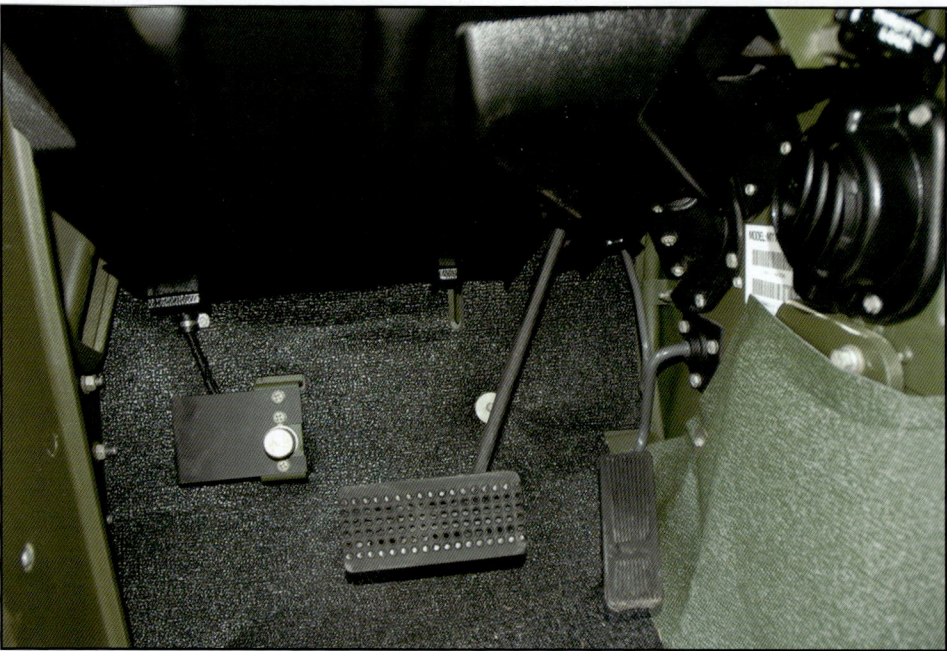

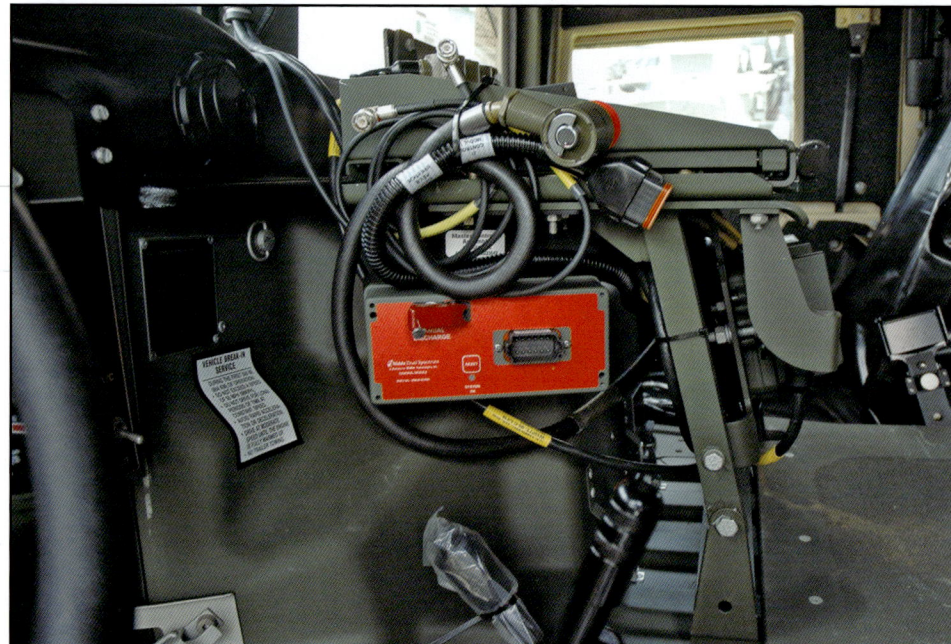

Top left: On the left side of the dashboard is the starter switch and, below it, the light switches. The directional signal lever is between those switches and the steering wheel. The placard next to the air conditioner vent has instructions on adjusting the driver's seat. **Top right:** To the front of the driver's seat are, left to right, the dimmer foot switch, the brake pedal, and the accelerator pedal. Thermal blanket material is applied to the floor and the interior tunnel (right) for insulation purposes. To the top right is the hand throttle. **Above left:** Below the instrument panel, at the lower center, is the fording switch. Before the vehicle enters deep water, the switch is turned to the "deep ford" position indicated at the bottom of the switch plate. Once the vehicle is out of deep water, the switch is turned to "vent." **Above right:** The engine-access cover is viewed from the left side. The red box mounted on its side is part of an automatic fire-suppression system. Mounted over the access cover is the radio rack, with its supports and a microphone stand to the rear of the access cover.

Top left: In this factory-fresh M1151A1 w/B1, protective clear plastic wrap is applied around the visors. At the center is a tachometer. Just beyond it is the windshield washer motor. To the right is the front of the ring of the armament weapon station. **Top right:** The tachometer and windshield washer motor are viewed from the right front of the cab. The yellow caution sign on the windshield frame advises that high-intensity-noise hearing protection is required: the interior of a HMMWV can be a very noisy place. **Above left:** The engine access cover and the radio rack are viewed from the right side of the cab. At the top of the dashboard in the foreground are two round air-conditioner vents. The yellow decal below the right vent contains instructions on vehicle tie-down procedures. **Above right:** Below the rear of the radio rack are communications controls. To the left is an intercom control box. To the right is the master control station/light, which provides interfaces for the vehicle power supply and connection of two onboard radios.

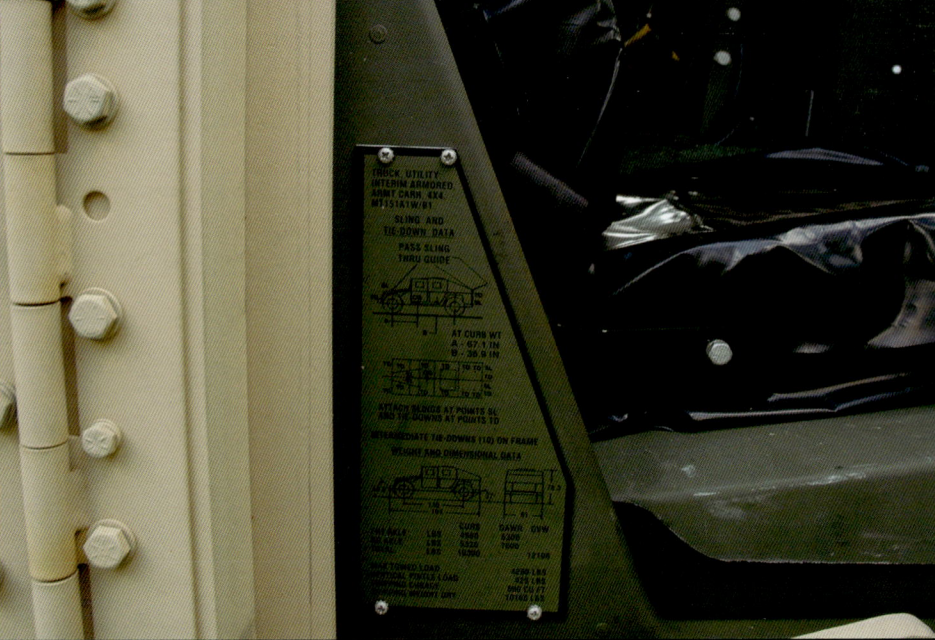

Top left: The rear of the crew compartment of a new M1151A1 w/B1 is viewed from the front right of the cab. The seats are temporarily covered with protective plastic. The C-pillar bulkhead to the rear has two sliding doors for access to the cargo compartment. **Top right:** Affixed to the lower part of the B-pillar next to the companion seat is a placard specific to the M1151A1 w/B1 giving instructions on attaching an airlift sling or tying-down the vehicle. A close-up view is offered of a piano-type crew door hinge to the left. **Above left:** The interior of a crew door in an M1151A1 w/B1 is shown. Crash pads applied to the door have cutouts for the handle and the soft handgrip. To open the door from inside the cab, the handle is pulled upward; the handle is pushed down to open it from the outside. **Above right:** The front right crew door is open, showing details of the window. The ball-shaped object is a handle/latch for the sliding window; a detent hole is in the sill. The doors on the M1151A1 w/B1 are part of the Frag 5 armor kit, designed to withstand roadside bombs.

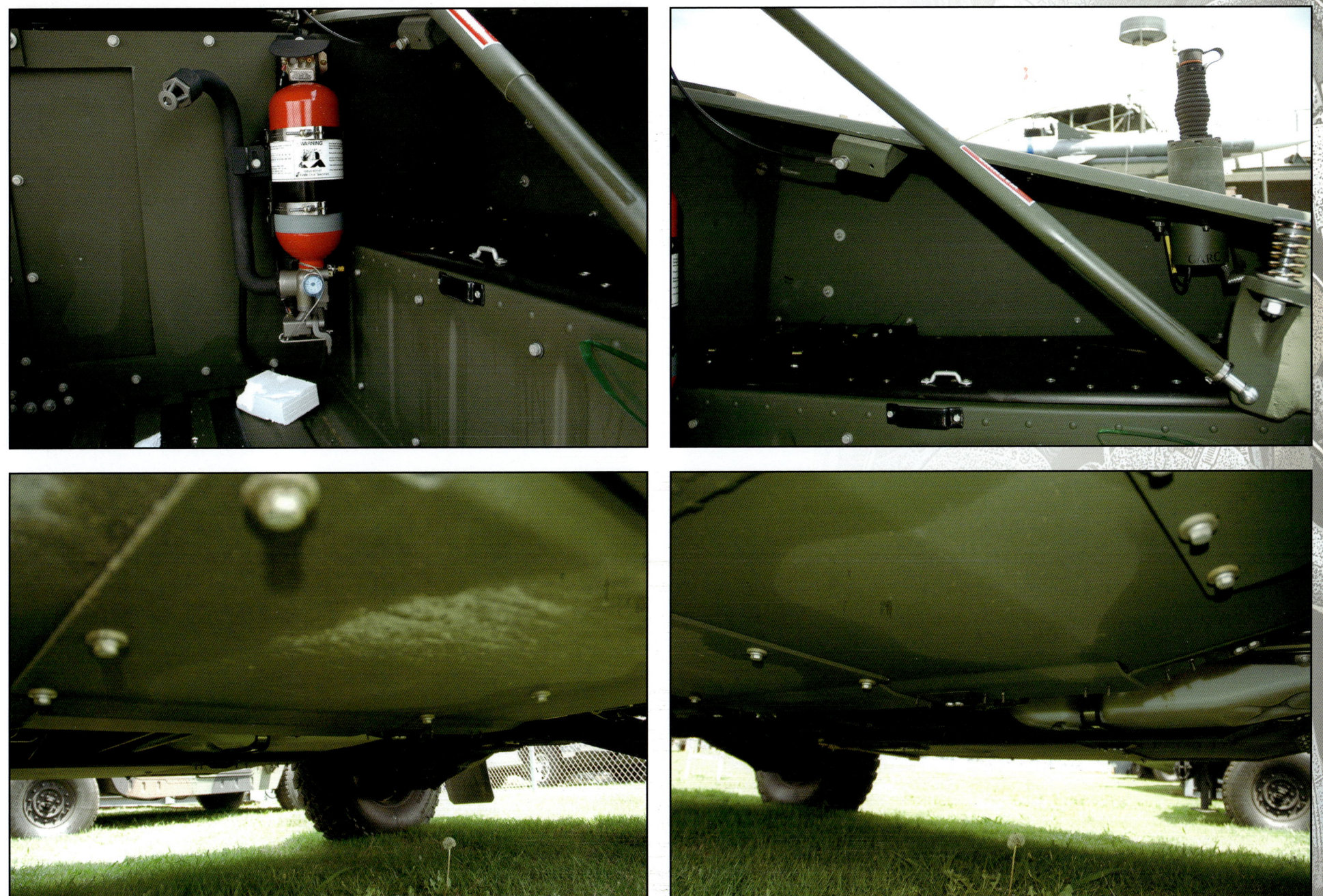

Top left: In forward right side of the cargo compartment of the M1151A1 w/B1 is a fire extinguisher. To the left is the C-pillar door, positioned in the ballistic bulkhead at the rear of the cab. The right rear wheel house is to the right. **Top right:** This view is into the right side of the cargo compartment, showing the space above the wheel house. The tubular assembly running diagonally across the photo is the door closer. The base of the radio antenna matching unit/base protrudes below the cargo shell. **Above left:** A look underneath the M1151A1 w/B1 reveals armor plates applied in some places below the frame. Like the rocker armor and the energy-absorbing seats, under-body armor on the M1151A1 w/B1 is part of the vehicle's stock armor-protection suite. **Above right:** Another view of the armor under the chassis of the M1151A1 w/B1 shows that the fuel tank to the right of the photo does not have armor underneath it. The underbody armor is intended primarily to protect the crew from mine explosions.

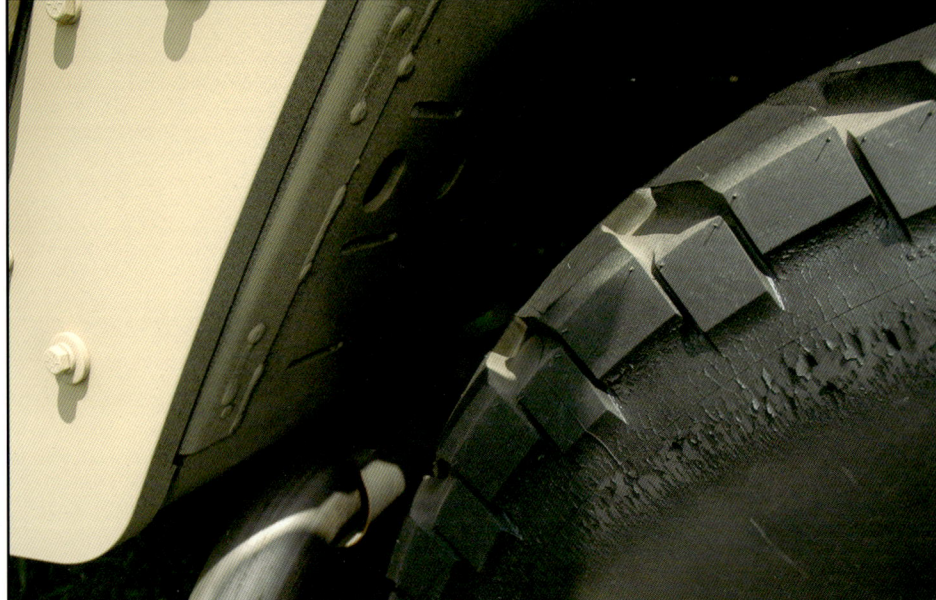

Top left: The mounting straps of the fuel tank are visible. To the upper right is the tail pipe of the exhaust system. When the vehicle is readied for deep-water fording, the tail pipe is removed from the muffler and an extended tail pipe is installed in its place. **Top right:** The rear differential, as viewed from the rear of the M1151A1 w/B1, is framed by the rear cross member of the frame, at the bottom corners of which are the rear mounting brackets for the lower A-arms of the rear suspension. **Above left:** The tailpipe (bottom center) and its hanger are viewed from the upper front of the left rear tire. A small part of the muffler is visible on the other side of the frame. To the right of center is the upper A-arm and its mounting bracket. **Above right:** The design of the front panel of the left rear wheel house, or wheel well, is displayed. At the same time, the tread pattern on the side of the B. F. Goodrich Baja T/A 37-12.5, load-range D tire is shown. At the bottom is the tail pipe.

M1165A1

A factory-new M1165A1 w/B3 is on display with its hood open. This HMMWV expanded-capacity command-and-control and general-purpose vehicle presents an appearance quite similar to that of the M1151A1 w/B1 in the preceding series of photos, but without the cargo shell on the rear of the body. In addition to the factory-installed underbody armor, rocker armor, lower windscreen deflector armor, and energy absorbing seats common to the M1165, this version also has the perimeter armor, roof armor, and Frag 5 doors.

Top left: This is a driver's-eye view of the cab of a brand-new M1165A1 w/B3, facing toward the right rear corner and showing several energy-absorbing seats. At the top is the filled-in opening for an optional weapon armament station. To the right is a fire extinguisher. **Top right:** A closer view is provided of the rear of the cab, where the top of the C-pillar ballistic bulkhead meets the ceiling. At the center, just aft of the opening for the weapon armament station, is an intercom control box for personnel in the rear seats. **Above left:** The intercom control box at the upper rear of the cab is viewed close-up. This unit includes separate selectors for volume and mode for two headsets. Around the volume controls are color-coded rings, with red indicating volumes that may induce hearing loss. **Above right:** On top of the interior tunnel toward the rear of the cab is a console with air-conditioner vents. Air conditioning is essential in up-armored HMMWVs where crewmen could otherwise suffer from the debilitating effects of prolonged exposure to heat.

This M1165A1 w/B3 displays an open cargo compartment at the rear. The rear of the C-pillar ballistic bulkhead is visible. Mounted on the rear of the body are two radio antenna matching unit/bases. An air conditioner grille is over the rear wheel house.

Although this M1165A1 w/B3 displays U.S. Marine Corps markings, the rear bumper is the wide version, not the narrow version characteristic of earlier USMC HMMWVs. Mud flaps are fitted on the bumpers, and tail composite lights and reflectors are on the body.

HMMWV General Data

Model	M997A2	M1025A2	M1035A2	M1043A2	M1097A2
Weight Net (pounds)	7,168	6,780	6,100	7,264	5,900
Weight Gross (pounds)	10,300	10,300	10,300	10,300	10,300
Length (inches)	204.5	190.5	182.5	190.5	190.5
Wheelbase (inches)	130	130	130	130	130
Width (inches)	86	86	86	86	86
Height (inches)	102	76	72	76	74
Track (inches)	71.6	71.6	71.6	71.6	71.6
Tire Size	37x12.5 R16.5	37x12.5 R16.5	37x12.5 R16.5	37x12.5 R16.5	37x12.5 R16.5
Max Speed (MPH)	70	70	70	70	70
Fuel Capacity (gallons)	25	25	25	25	25
Electrical	24-volt negative	24-volt negative	24-volt negative	24-volt negative	24-volt negative
Transmission Speeds	4	4	4	4	4
Transfer Speeds	2	2	2	2	2
Turning Radius (feet)	25	25	25	25	25

*Winch-equipped models have payloads reduced by 127 pounds.

Left: Over the left rear wheel house of the M1165A1 w/B3 is an air conditioner grille similar to the one over the right rear wheel house. Details of the perimeter armor around the C-pillar at the rear of the cab are in view to the left of the photograph.

Right: The side perimeter armor ends at the rear of the cab. What appears to be an armor plate is fastened to the top of each of the rear wheel houses, or wheel wells, most likely to provide protection to the air-conditioning equipment at the top of the wheel houses.

M1152A1 with B2 SECM

Several M1152A1 w/B2 Shop-Equipment Contact-Maintenance (SECM) trucks are lined up. These are up-armored cargo/troop carriers converted to mobile field repair stations, equipped for performing repairs that do not require a damaged vehicle to go to a fully equipped maintenance facility. They carry an array of hand and pneumatic tools, welding equipment, a cutting torch, an air compressor, spotlights, and other essential equipment, and can supply electrical power through the vehicle's power inverter. The version with the B2 armor kit provides armor protection from the rear of the front seats to the front of the cab.

The enclosure on the rear of the M1152A1 w/B2 SECM has roll-up side doors providing access to storage cabinets and drawers for the tools. All tools are stored in specific locations for ease of accessing them. Lifting eyes are visible on top of the enclosure.

Top left: On the rear of the SECM enclosure is a roll-up door. Three data plates are affixed to the right side of the enclosure. Below the enclosure on the rear of the body are a tailgate and a wide-type bumper with a tow pintle and a lifting shackle at each end of the bumper. **Top right:** Another roll-up door is on the left side of the SECM enclosure. The rear crew doors on these vehicles remain operable, to provide access to stowage and workspace inside. The driver's side ballistic window and frame protrudes from the front door. **Above left:** An M1152A1 w/B2 SECM is viewed close-up from the left side, showing the roll-up door on the side of the enclosure. At the bottom of the door is a sill with three brackets to hold a tube that functions as the handle for raising and lowering the door. (James Alexander) **Above right:** Before the M1152A1 w/B2 SECM, the SECM enclosure was carried on the M1113 Extended-Capacity Vehicle (ECV). Here, mechanics work from a different design of SECM on an HMMWV that appears to lack armor, providing a view of the tool cabinets.

Field use

As the HMMWV was increasingly pushed into the role of combat vehicle in the Middle East, efforts were made to create a variant better suited for this duty, with increased armor protection, and more power. The result was the M1151, shown here ready for shipment from AM General's South Bend, Indiana, facility. (AM General)

A winch-equipped HMMWV with an unusual antenna to the front of the left side of the windshield is part of a column moving through a busy town. To the rear is an uparmored HMMWV armament carrier with shields for the .50-caliber machine gunner. (Department of Defense)

The crew of an uparmored HMMWV armament carrier pause during a patrol. This vehicle is equipped with an Ibis Tek heavy-push front bumper and brush guard. A slot for a winch cable is at the center of the bumper. The gunner has the protection of an armored shield on three sides of the turret and a gun shield. (Department of Defense)

Troopers of the 82nd Airborne Division take a break next to their HMMWVs. Supplemental armor is fastened to the armor of the turret of the farther vehicle. Mounted on the front corners of its roof are four-tube smoke-grenade dischargers with dust covers; another discharger is to the rear of the roof on the right cargo shell. (Department of Defense)

A member of an HMMWV crew of the 82nd Airborne Division calls in a report while the gunner stands at the ready on the .50-caliber machine gun. From this angle, with the doors open, a good view is available of the armor plates applied to the outsides of the doors and the cutouts for the door handles. Note the dark-colored rear-view mirror and frame. (Department of Defense)

A gunner mans a .50-caliber machine gun on a HMMWV uparmored armament carrier. The low armored shield on the sides and rear of the weapons turret was designed to give the gunner some protection and maintain his all-around visibility. In the threat-intensive combat environment of Iraq and Afghanistan, it is easy to see why the troops in the field were clamoring for more and better armored protection for HMMWVs, and why they resorted to expedient fixes in the form of supplemental bolt-on armor. (Department of Defense)

Top left: Even an uparmored HMMWV was no match for a mine, an antitank rocket, or a roadside improvised explosive device, as the crew of this vehicle discovered. The engine compartment was destroyed, the hood and right front wheel blown off, and the left front wheel flattened. This vehicle had four smoke-grenade dischargers. **Top right:** Another uparmored HMMWV armament carrier suffered a hit to the engine compartment. This vehicle has "MILITARY POLICE" stenciled on the side of the cargo shell. **Above left:** The same HMMWV armament carrier is viewed from the front, showing the badly fractured hood. Markings on the fenders identify this as the number-83 vehicle of Headquarters, 42nd Military Police Brigade. **Above right:** All too many HMMWVs in Iraq and Afghanistan suffered the fate of this vehicle, which was totally destroyed by an explosion. It was this epidemic of attacks against U.S. convoy and patrol vehicles that led to the initiative in which the armed forces were supplied with new mine-resistant, ambush-protected (MRAP) vehicles that were better equipped to withstand mines and improvised explosive devices (IEDs). (Department of Defense, all)

Top left: An uparmored HMMWV armament carrier devastated by an explosion is being transported on an M3 flat rack of an Oshkosh M1075 PLS in Iraq. The hood of the HMMWV is gone, and the cargo shell door has been blown off. Above the roof is a weapons mount and shield. **Top right:** The rear end of this HMMWV armament carrier bore the brunt of an explosion, tearing off the right rear wheel and much of the bodywork and savaging the chassis frame. Faintly visible inside the torn-up tire lying on the ground is the run-flat insert, a two-piece magnesium disc that allows the HMMWV to continue driving even if the tire is punctured. **Above left:** A blast ripped the roof of this HMMWV, tore off the rear door, and twisted the front passenger's door. There also are several punctures in the right cargo shell. **Above right:** This uparmored HMMWV armament carrier equipped with shields for its turret and smoke-discharger arrays was totaled in a roadside blast. Mounted on the rear of the vehicle is a stowage rack made of tubular metal. (Department of Defense, all)

Top left: A number of interesting modifications are present on this uparmored HMMWV armament carrier. On the front end is a tubular bumper with a brush guard on top fashioned from metal slats on a strap-metal frame. A dark-colored scoop has been installed over the hood louvers. What appears to be a small loudspeaker is mounted on the right A-column. **Top right:** A line of HMMWVs bristling with machine guns and antennas is ready for action in the desert. The first vehicle is armed with an M240-family 7.62mm machine gun, while the rest of the vehicles are armed with .50-caliber machine guns. Note the two ammunition boxes on the side shield of the second HMMWV. **Above left**: Parked along an Iraqi highway, this uparmored HMMWV armament carrier bears markings for the 3rd Squadron, 3rd Cavalry Regiment. It is armed with a .50-caliber machine gun and is equipped with four four-tube smoke-grenade dischargers with dust covers. Stashed behind the brush guard on the heavy-push bumper is a tow bar. **Above right:** The crew of an uparmored HMMWV with "DX" marked on the rear door confers at an overlook above a desert base. A spare tire is secured to the rear of the vehicle with webbing straps. The HMMWV is armed with a .50-caliber machine gun, with an armor kit installed on the turret. Smoke-grenade dischargers are present, and an AT4 recoilless antitank weapon is strapped to the rear of the shield. (Department of Defense, all)

An uparmored HMMWV cargo/cargo carrier is parked next to an M1075 PLS at a desert base. A rather beat-up sheet-metal air scoop has been installed over the hood louvers. In an unusual touch, a pioneer tool rack is mounted on the soft cover of the cab; part of a shovel is visible on the rack. (Department of Defense)

Marine Lance Corporal Michael Winniford of Alpha Company, 8th Engineer Support Battalion, has left his uparmored M1116 HMMWV armament carrier to inspect a pothole for the presence of a mine on Main Supply Route Tin during Operation Iraqi Freedom. He was a member of the Engineer's Sapper Team, which cleared land mines from the highways in advance of convoys. (Department of Defense)

Several USMC HMMWV cargo/personnel carriers of 1st and 2nd Platoons, Echo Company, 2nd Battalion, 5th Marines, are lined up, part of a supply convoy to a women's and children's hospital in Ar Ramadi during Operation Iraqi Freedom. The nearest HMMWV is a motley-looking vehicle, sporting cannibalized parts and improvised armor. Partly hidden behind the HMMWV to the left is an HMMWV armament carrier. (Department of Defense)

US Marine Corps (USMC) First Lieutenant (1LT) Edward Orillion (left), 22nd Marine Expeditionary Unit (MEU) Special Operations Capable (SOC), pulls security duty with his Colt 5.56mm M16A2 Assault Rifle during a brief convoy halt while conducting an overt vehicular reconnaissance patrol throughout the Kandahar and Oruzgan Provinces of Afghanistan (AFG) during Operation ULYSSES II, which is the first combat operation undertaken by the 22nd MEU (SOC) during Operation ENDURING FREEDOM. (Department of Defense)

With a vast assortment of weaponry at the ready, US Marine Corps (USMC) personnel from Battalion Landing Team (BLT), 1st Battalion, 6th Marines and the ground combat element of the 22nd Marine Expeditionary Unit (MEU), wait in their High-Mobility Multipurpose Wheeled Vehicles (HMMWV) for the word to move toward a mountain where Taliban snipers are engaging their convoy, near the village of Siah Chub Kalay, Afghanistan (AFG) during Operation ASBURY PARK and Operation ENDURING FREEDOM. The improvised side armor on these vehicles is of interest. (Department of Defense)